Lecture Notes in Mathematics

Edited by A. Dold and B. Eckmann

1163

Iteration Theory and its Functional Equations

Proceedings of the International Symposium
held at Schloss Hofen (Lochau), Austria
Sept. 28–Oct. 1, 1984

Edited by R. Liedl, L. Reich and Gy. Targonski

Springer-Verlag
Berlin Heidelberg New York Tokyo

Editors

Roman Liedl
Institut für Mathematik, Universität Innsbruck
Innrain 52, 6020 Innsbruck, Austria

Ludwig Reich
Institut für Mathematik, Universität Graz
Brandhofgasse 18, 8010 Graz, Austria

György Targonski
Fachbereich Mathematik, Universität Marburg, Lahnberge
3550 Marburg, Federal Republic of Germany

Mathematics Subject Classification (1980): 39-xx, 58Fxx, 70Kxx, 92-xx

ISBN 3-540-16067-1 Springer-Verlag Berlin Heidelberg New York Tokyo
ISBN 0-387-16067-1 Springer-Verlag New York Heidelberg Berlin Tokyo

Printing and binding: Beltz Offsetdruck, Hemsbach/Bergstr.
2146/3140-543210

INTRODUCTION

These are the Proceedings of the International Symposium on Iteration Theory and its Functional Equations held at Schloss Hofen in Lochau, (Vorarlberg, Austria), Sept. 28 to Oct. 1, 1984.

It was the fifth in a series of international meetings on iteration theory, following Toulouse (1973), Schloss Retzhof near Graz (1977), Amöneburg near Marburg (1980) and again Toulouse (1982). We hope there will be further meetings.

Many of the papers are of the convential "definition-theorem-proof" type of mathematics. Several others could be thought of as belonging to "experimental mathematics". The term is a controversial one and we do not wish to imply that all, or even that some, of the authors in question would classify their work under experimental mathematics. We personally feel there is nothing wrong with this term. Used by P.R. Stein and S. Ulam as early as 1964 it means numerical work, usually with a graphic display of the results, and formulating conjectures based on these numerical and graphic results, often also furnishing a rigorous mathematical proof.

One paper (Rössler's) ventures into "no man's land" between mathematics, physics and philosophy, offering stimulating heuristic ideas.

The overall impact of the symposium was summed up in a "round-table discussion" held on the evening of the last day.

The fundamental question, of course, was "What is iteration theory?" and it was answered implicitly by placing it between two old-established fields of mathematics: Functional Equations and Dynamics. This means for us close co-operation with colleagues working in these fields. But this co-operation, in turn, is fraught with grave problems concerning terminology and notation. Consider the term "orbit". For us, it means the set of all predecessors of all successors of an element under a given mapping; it can be quite complicated a structure and the notion is basic for iteration theory. For the dynamicist, on the other hand, orbit means the sequence of successors of a given element. This we call iteration sequence or, using a term introduced by the logicians, splinter. A misunderstanding of the term "orbit" induced a referee, a few years ago, to reject a brilliant paper by a young iteration theorist; the paper was subsequently published by a prestigious journal.

We will not, of course, abandon the terminology Gordon Whyburn gave us more than four decades ago; but we could speak of Kuratowski-Whyburn orbits (KW-orbits for short) whenever the danger of confusion arises.

An important idea was contributed by Serge Dubuc, illustrated on the example of his own mathematical development. Iteration theory is not only "une voie royale pour une majorité de questions en analyse numérique et en optimisation" but also, and for us primarily, subject of interest in its own right. In this context, the theory of fractals comes to the mind, and also the multitude of beautiful, nowadays even coloured, patterns for which the term "chaos esthétique" was coined by Igor Gumowski and Christian Mira.

The round-table conference also served as a survey of work done since 1973, when the first meeting devoted to iteration theory took place in Toulouse. The study of embeddings (analytic, continuous, or without regularity conditions) of formal power series (in Graz) the discovery of the Pilgerschritt method for continuous iteration (in Innsbruck), the solution of the general problem of iterative roots (in Marburg), the full solution of the embedding problem for continuous functions of one real variable (in Katowice) and the revival of the French school of complex iteration, also the continuing interest in strange attractors as well as the application of resurgent functions to the iteration of local analytic diffeomorphisms (in Orsay), were among the themes mentioned with some satisfaction. At least one important question was answered. At the third meeting (Amöneburg 1980) the question was asked (also at a round table discussion) whether "massive" chaos exists, that is, whether continuous mappings of a real interval into itself exist such that, with the exception of a set of Lebesque measure zero, all points belong to the "scrambled set". At this meeting, Michał Misiurewicz gave an affirmative answer in presenting his paper "Chaos almost everywhere".

The Vorarlberger Landesregierung (State Government of Vorarlberg) and the staff of the Landesbildungszentrum Schloss Hofen (Continued Education Centre of Vorarlberg State at Hofen Castle) helped us to make the meeting a pleasant one from every point of view.

Our sincere thanks go to Dr. Norbert Netzer. He was involved in the preparations from the beginning, planned and prepared all the details, and ensured smooth functioning of the program during the symposium. He also took part in the preparation of these Proceedings.

We are very much indebted to Professor Detlef Gronau for his untiring work during the preparation of the Proceedings.

We wish to express our gratitude to the Österreichische Forschungsgemeinschaft, Vienna, the Vorarlberger Landesregierung, Bregenz, and the Bundesministerium für Wissenschaft und Forschung, Vienna, for most generous financial support.

Our thanks are due to Mrs. Maria Krenn for an excellent typing job.

We hope the reader will agree that the symposium was a step forward for our growing field of iteration theory.

Innsbruck, Graz and Marburg, June 1985

Roman Liedl
Ludwig Reich
György Targonski

CONTENTS

ON SOME PROPERTIES OF AN ABSORPTIVE AREA AND A CHAOTIC AREA FOR AN R^2-ENDOMORPHISM

A. Barugola

1. INTRODUCTION

Take the non-invertible recurrence or point mapping T:

$$x_1 = f(x,y,\lambda) , \qquad y_1 = g(x,y,\lambda) , \tag{1}$$

where f and g are single-valued functions that are continuous with respect to the real variables x,y and the real parameter λ and are continuously differentiable at least once with respect to their arguments; T is an endomorphism, the Jacobian of which vanishes on a curve of the plane (x,y), for which we write (LC_{-1}).

Systems of this type have been found to involve complex dynamical behaviour, and have recently been shown to include absorptive areas and chaotic areas [1-9]. A general method for obtaining these areas has been described [2,6]. The terminology and notation adopted below will conform to the definitions given in [1,2,3].

The aim of this paper is to describe some results obtained at our laboratory since the 'Colloque sur la Théorie de l'Itération...' which was held in Toulouse in May 1982.

2. A PROPERTY OF AN ABSORPTIVE AREA

By applying to a wide range of examples the method for determining an absorptive area (d') or a chaotic area (d) described in [2,6], it has been possible on the one hand to classify the various cases obtained, and on the other hand, to describe some new bifurcations occurring in the absorptive and chaotic areas [11], particularly the bifurcation which leads from a chaotic area having K non-connected domains, $K \geq 2$, to one having a single connected domain [10].

An important class of mappings T has been found to exist in which the following hypotheses are satisfied: 'T admits an absorptive area that is connected and invariant under transformation by T'. Several properties have been established for this class of mappings [11]. The most significant of these is as follows:

Proposition 1. The boundary (f') of an absorptive area (d') that is connected and invariant under transformation T contains at least one segment of (LC), the critical line of T; and (d') contains at least one antecedent of this segment, located on the curve (LC_{-1}); (LC) = T((LC_{-1})).

3. ANNULAR CHAOTIC AREA

In the general case, $(d) \subseteq (d')$; when $(d) \subset (d')$, (d) is annular if it complies with the following description:

- (d'), delimited by its boundary (f'), is simply connected;
- the sequence of iterates of any point $M, M \in (D) - (d')$ enters (d') and subsequently remains enclosed therein (D) being the domain of influence or attraction of (d') limited by a boundary (F));
- the sequence of iterates of any point $M, M \in (W) = (d') - (d)$, ($M \neq P$, if a repulsive fixed point P is included within (W) enters (d) and subsequently remains enclosed therein.

(W) is a simply connected open set, called a 'hole'; the boundary (f) of (d) is such that $(f) = (f_e) \cup (f_i)$, $(f_e) \cap (f_i) = \Phi$, where (f_e) is the outer boundary and (f_i) the inner boundary. In general $(f_e) = (f')$, $T((d')) = (d')$, and $T((d')-(W)) \subseteq (d')-(W)$. Moreover, (f'), (f_e) and (f_i) consist of segments of critical lines which it is possible to determine for a given mapping T.

The existence of a hole in a chaotic area has already been observed in numerical studies (see for example [4] figure 1, [7] figure 1).

4. ANALYSIS OF THE EXISTENCE OF A HOLE

In the presence of a hole (W), two cases can arise, according to whether or not it contains a repulsive fixed point. Only the former case will be considered here: the repulsive fixed point is a node or a focus, and cannot be a saddle point, since if it were, the invariant curve ω corresponding to the multiplier having a modulus smaller than one would give the saddle point an attractive fixed point for the sequence of iterates restricted to ω. For a different set of λ values, the repulsive fixed point contained in (W) constitutes an attractive fixed point, from which one can generally obtain the chaotic behaviour of the solution(s) contained in (d) through various types of bifurcations.

For the sake of simplicity, we shall consider only the following case here: (W) contains a repulsive fixed point $P, T((d')) = (d'), T((d') - (W)) = (d') - (W)$; moreover, (W) is entirely situated on the one hand in a region having a constant number of antecedents, and on the other hand, in a region for which the sign of the Jacobian of the right-hand side of (1) is constant (in which case (LC_{-1}) does not cross (W)). (LC), the critical line of T, divides the plane (x,y) into two regions, $\mathbb{Z}(p)$ and $\mathbb{Z}(q)$, having p and q rank one antecedents: $p-q = 2h$, $h = 1,2,\ldots, q = 0,1,\ldots$ These hypotheses lead to the following proposition:

Proposition 2. If the hole (W) $\subset \mathbb{Z}(p)$ (or $\mathbb{Z}(q)$, $q \neq 0$),(W) has an antecedent that is contained within (W) and that itself contains the fixed point P and (p-1) (or (q-1)) disjoint antecedents located outside (d'), each containing an antecedent of P.

The proof of this proposition is based on the procedure described in [2,3]:
- (W) cannot belong to $\mathbb{Z}(q)$ when q = 0 according to property P_7 as defined in [2,3];
- if (W) $\subset \mathbb{Z}$ (p) (or $\mathbb{Z}(q)$, $q \neq 0$),(W) will have p (or q) disjoint antecedents $(W_{-1})^i$, i = 1 to p (or i = 1 to q) and each of these antecedents will contain an antecedent $(P_{-1})^i$ of the fixed point P. Since P is invariant under transformation by T and T^{-1}, one of the antecedents of P is P itself and hence there is an antecedent of (W) included within (W), while the other antecedents are outside (d');
- if (W) $\subset \mathbb{Z}(q)$ when q = 1, (W) has only one antecedent, and it is included in (W).

5. BIFURCATION PRODUCING OR ABOLISHING A HOLE

If for a given system (1), we take the direction in which parameter λ varies to be a growing one, for example, it is possible in certain cases to determine the bifurcation producing a hole in a chaotic area, as well as the bifurcation causing the hole to disappear. Production of a hole is generally obtained by the bifurcation leading from a chaotic area having K non-connected domains to a chaotic area having a single connected domain [10].

The following example is given to describe a bifurcation causing a hole to disappear. Consider the mapping:

$$x_1 = 3.9x(1-x-y), \qquad y_1 = \lambda xy, \qquad x > 0, \qquad y > 0, \qquad \lambda > 0. \tag{2}$$

The domain of influence (D) of the absorptive area (d') is the region: x > 0, y > 0, 1-x-y > 0. The equation expressing (LC_{-1}) is x = 0.5, and (LC) separates two regions having two and zero rank one antecedents. A continuous set of values for λ exists for which the chaotic area is annular and for which (d'), (d) and (W) are such that T((d')) = (d'), and T((d') - (W)) = (d') - (W).

Figure 1, which corresponds to λ = 3.23, shows areas (d') and (d), the hole (W), and the two antecedents of the latter, $(W_{-1})^i$, i = 1,2. (d') consist of a simply connected domain, the boundary (f') of which is composed of segments of critical lines (LC), and $(LC_j) = T^j(LC)$, j = 1 to 5:b_1a_1 of (LC),a_ja_{j+1} of (LC_j) when j = 1,2,3,a_4d_0 of (LC_4),b_1a_5 of (LC_4) and a_5d_0 of (LC_5). (In figures 1 and 4, the (LC_j) are simply written L_j). The annular chaotic area (d) is delimited by a boundary (f) = $(f_e) \cup (f_i)$, (f_e) = (f') and (f_i) being composed of g_0e_0 of (LC_5),e_je_{j+1} of (LC_{6+j}),j = 0,1,2 and e_3g_0 of (LC_9), where $b_1 = T(b_0)$, $a_j = T^j(a_0)$ and e_j =

FIGURES

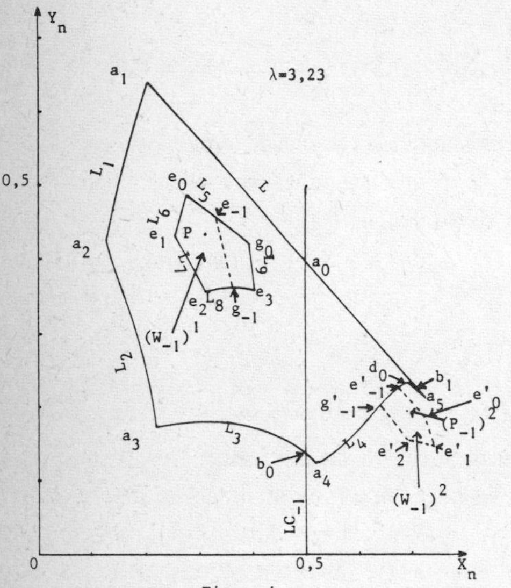

Figure 1.

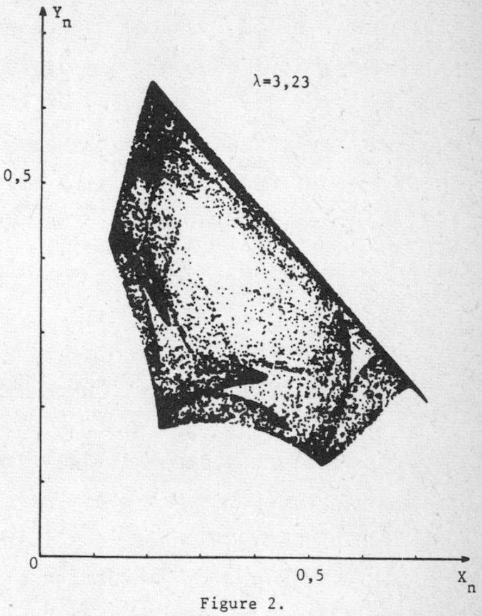

Figure 2.

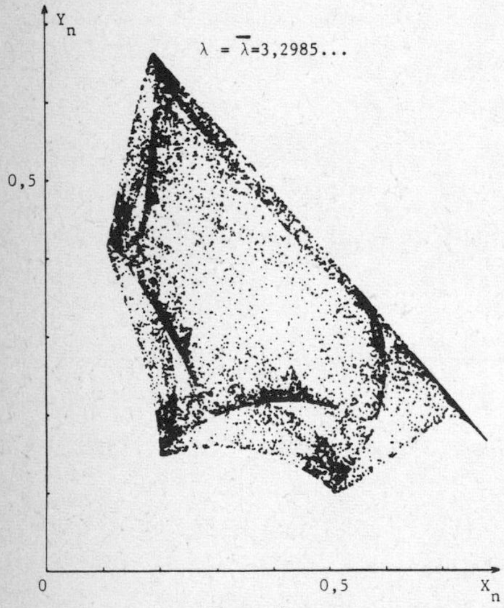

Figure 3.

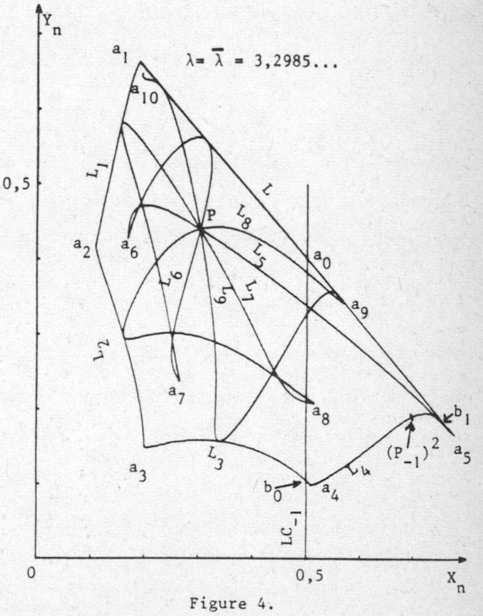

Figure 4.

$T^j(e_0)$. The segments of critical lines have contact points of order one at b_1 and a_j, $j = 1$ to 5, whereas at d_0, g_0 and e_j, $j = 0$ to 3, the contact points are of order zero. P is the repulsive focus fixed point and $(P_{-1})^1 = P$ and $(P_{-1})^2$ its two antecedents; likewise, $(W_{-1})^1$ is the antecedent of (W) containing P and $(W_{-1})^2$ the antecedent of (W) that is not included in (d') and which contains $(P_{-1})^2$.

Figure 2 illustrates a numerical test through which the hole (W) can be shown.

When λ increases, $\lambda < \bar{\lambda} = 3.2985...$,(W) and its two antecedents decrease in area, and $(P_{-1})^2$ tends towards (d'). When $\lambda = \bar{\lambda}$ (fig.3), (W) disappears, and $(W_{-1})^2$ is reduced to $(P_{-1})^2$, which is located on the segment a_4d_0 of (LC_4) and of (f'). When $\bar{\lambda} < \lambda < 4$, the chaotic area (d) = (d'), and $(P_{-1})^2$ is included within (d'). $\lambda = \bar{\lambda}$ is a bifurcation value corresponding to either the disappearance or the appearance of the hole (W) (according to the direction in which λ varies); with this value, $(P_{-1})^2$ is located on the segment a_4a_5 of (LC_4), from which it can be deduced that the successive iterates of all orders a_ja_{j+1} of (LC_j), $j = 5,6,...$, cross through the fixed point P (fig.4).

6. COMMENTS

The phenomenon described above in the case of a mapping T can be reproduced taking T^k, $k = 2,3,...$, and one can obtain k holes, each containing one point belonging to a repulsive cycle of the order k. A forthcoming paper [12] will present a more thorough investigation of the existence of holes in a chaotic area and describe their properties.

REFERENCES

[1] Gumowski, I., Mira, C.: Dynamique chaotique, Cépadues éd., Toulouse, 1980, Ch.6.

[2] Barugola, A.: Lignes critiques et zones absorbantes pour une récurrence du second ordre à inverse non unique. Colloque sur la Théorie de l'Itération et ses Applications, éd.du C.N.R.S., Toulouse, 17-22 Mai 1982, p.83.

[3] Barugola, A.: Quelques propriétés des lignes critiques d'une récurrence du second ordre à inverse non unique. Détermination d'une zone absorbante. R.A.I.R.O., Analyse Numérique, Vol.18, n°2, 1984, p.137.

[4] Gumowski, I., Mira, C.: Solutions chaotiques bornées d'une récurrence ou transformation ponctuelle du deuxième ordre, à inverse non unique. C.R.Acad.Sci. Paris, 285, Série A, 1977, p.477.

[5] Barugola, A.: Détermination de la frontière d'une zone absorbante relative à une récurrence du deuxième ordre, à inverse non unique. C.R.Acad.Sci.Paris, 290, Série B, 1980, p.257.

[6] Cathala, J.C.: Détermination de zones absorbantes et chaotiques pour un endomorphisme d'ordre deux. Colloque sur la Théorie de l'Itération et ses Applications, éd. du C.N.R.S., Toulouse, 17-22 Mai 1982, p.91.

[7] Gumowski, I., Mira, C.: Bifurcation déstabilisant une solution chaotique d'un endomorphisme du deuxième ordre. C.R.Acad. Sci. Paris, 285, Série A, 1978, p.427.

[8] Cathala, J.C.: Sur la dynamique complexe et la détermination d'une zone absorbante pour un système à données échantillonnés décrit par une récurrence du deuxième ordre. R.A.I.R.O., Automatique, V.16, n°2, 1982, p.175.

[9] Cathala, J.C.: Absorptive area and chaotic area in two-dimensional endomorphisms. Non Linear Analysis, V.7, n°2, 1983, p.147.

[10] Cathala, J.C.: On the bifurcation between a chaotic area of T^K and a chaotic area of T.Communication at this Symposium.

[11] Barugola, A.: Sur certaines zones absorbantes et zones chaotiques d'un endomorphisme bidimensionnel. Non Linear Mechanics (to appear).

[12] Barugola, A., Cathala, J.C., and Mira, C.: (to appear).

Université de Provence
Département d'Automatique
et de Dynamique Non Linéaire
Rue H. Poincaré
13397 Marseille Cedex 13, France

A FUNCTIONAL EQUATION FOR THE EMBEDDING OF A HOMEOMORPHISM OF THE INTERVAL INTO A FLOW

W.A.Beyer and P.J.Channell*)

ABSTRACT

A functional equation $\phi(\omega(x)) = \omega'(x)\phi(x)$ for $\phi(x)$ is derived for the problem of finding ϕ so that for a given orientation-preserving everywhere-differentiable homeomorphism $\omega(x)$ of $[0,1]$ with $\omega' \neq 0$, there exists a solution $F(x,t)$ to $F_t(x,t) = \phi(F(x,t))$ so that $F(x,0) = x$, $F(x,1) = \omega(x)$. A solution to the functional equation is given for the case where ω has a finite number of fixed points a_i with $\omega'(a_i) \neq 1$. The analogous equation in n-dimensional space is given.

1. INTRODUCTION

Suppose one has a homeomorphism $\omega(x)$ of $[0,1]$ which is everywhere differentiable and orientation-preserving with $\omega' \neq 0$ and has a finite number of fixed points $0 = a_0 < a_1 < a_2 < \ldots < a_n = 1$. Further, $\omega'(a_i) \neq 1$. One wishes to embed this mapping as follows. Given $\omega(x)$, find $\phi(y)$ such that the solution $F(x,t)$ to the initial value problem

$$\frac{\partial F(x,t)}{\partial t} = \phi(F(x,t)) \quad , \tag{1.1}$$

$$F(x,0) = x \quad , \tag{1.2}$$

exists for $t \in [0,1]$, $x \in [0,1]$, and

$$F(x,1) = \omega(x) \quad . \tag{1.3}$$

We impose the conditions on ϕ that

(a) $\phi(a_i) = 0$, (1.4)

(b) a_i are the only zeros of ϕ , (1.5)

(c) ϕ is continuous on $[0,1]$. (1.6)

---

*) Presented by W.A.Beyer

This problem is relevant to the study of functional iteration since it permits iteration to be considered in terms of a dynamical system. A recent survey of the problem is given by Utz (1981). Our interest is in the application of a functional equation to construct a ϕ which solves the problem.

Remarks.

(a) As Palis (1974) has shown, in a sense he makes precise, the above embedding problem is not "generically" solvable for compact C^∞ manifolds without boundary.

(b) The work of Zdun (1979) on the problem of embedding a continuous map of [0,1] into an iteration semigroup is reviewed by Targonski (1981, 1984). Our hypotheses on ω puts ω into Zdun's class HE and thus ω must be embeddable in a continuous iteration semigroup. See Targonski, 1984, page 32.

(c) Some of the work in the present paper follows from Zdun (1977). In Section 2 we derive a functional equation for $\phi(y)$ and in Section 3 we discuss its solution. In Section 4 we given an example.

2. FUNCTIONAL EQUATION FOR $\phi(x)$

Since we assume that the a_i are the only zeros of ϕ and that ϕ is continuous, any solution $F(x,t)$ for x fixed and $x \neq a_i$ is either strictly increasing or strictly decreasing.

The function ϕ which we construct will satisfy a Lipschitz condition on any closed subinterval W of (a_i, a_{i+1}). Thus, if the initial value x is interior to W and if the solution remains in W, the existence and uniqueness theory (see Coddington and Levinson (1955)) of ordinary differential equations applies. For $x = a_i$, we define the solution to be $F(a_i,t) = a_i$ for $t \geq 0$.

Theorem 1. Suppose $\phi(x)$ satisfies (1.4), (1.5) and (1.6) and that $F(x,t)$ defined by (1.1) and (1.2) also fulfills (1.3). Then $\phi(x)$ satisfies the functional equation

$$\phi(\omega(x)) = \omega'(x)\phi(x) \quad (x \neq a_i,\ i = 0,1,\ldots n.) \tag{2.1}$$

Proof. Fix $x \in (a_i, a_{i+1})$. Put $y(t) = F(x,t)$. Then (1.1) can be put in the form

$$y'(t) = \phi(y(t)) \quad . \tag{2.2}$$

The function $y(t)$ is invertible since $\phi \neq 0$ by (1.5). Put $u = y(t)$. Then (2.2) can be put in the form

$$y'(y^{-1}(u)) = \phi(u) \quad .$$

Hence

$$y^{-1}(z) - y^{-1}(z_0) = \int_{z_0}^{z} \frac{du}{y'(y^{-1}(u))} = \int_{z_0}^{z} \frac{du}{\phi(u)} = : G(z) \quad .$$

Put $z = y(t)$. Then

$$t - y^{-1}(z_0) = G(y(t)) \ .$$

Choosing $t = 0$, then since $y(0) = x$, we have

$$-y^{-1}(z_0) = G(x) \quad .$$

Hence,

$$t + G(x) = G(F(x,t)) \quad .$$

With $t = 1$ we have

$$1 + G(x) = G(\omega) = G(\omega(x)) \quad . \tag{2.3}$$

Differentiation of (2.3) gives (2.1).

Remarks.

(a) The functional equation (2.1) can be generalized to n dimensions as:

$$\phi(\omega(x)) = D\omega(x)\phi(x) \tag{2.4}$$

where $x \in R^n$, and $\omega,\phi : R^n \to R^n$ and $D\omega$ is the Jacobian matrix. We will not give the proof here.

(b) McKiernan (1963) obtained (2.1) in his work on convergence of a series of iterates. Let $\omega(z)$ be analytic near $z = 0$ with the rth iterate $\omega^r(x)$ in the domain of ω for $r = 1,2,\ldots$. He then shows that the series

$$\phi(z) = \sum_{n=1}^{\infty} \frac{(-1)^{n+1}}{n} \sum_{r=0}^{n} \binom{n}{r} (-1)^{n-r} \omega^r(z) \tag{2.5}$$

converges in some neighbourhood of $z = 0$ provided $\omega(0) = 0$ and $\omega'(0) = \alpha$ for real $\alpha \varepsilon (0,1)$ and $\phi(z)$ satisfies $\phi(\omega(x)) = \omega'(x)\phi(z)$. It would be of interest to know if (2.5) solves the embedding problem.

(c) Jabotinsky also derives an equation which reduces to (2.1). Targonski (1984) calls that equation Jabotinsky's third equation.

(d) The functional equation (2.1) is of linear type and is discussed in Kuczma (1968), especially Chapters 2 and 4. If one takes logarithms of both sides of (2.1), one obtains a form of a linear functional equation discussed by Coifman and Kuczma (1969).

3. SOLUTION OF THE FUNCTIONAL EQUATION FOR ϕ

Our derivation of a solution follows that in Kuczma (1968). We suppose x is in the interval (a_i, a_{i+1}) and $\omega(x) > x$ on the interval. (The proof is similar in the contrary case that $\omega(x) < x$). We wish to find a solution to (2.1) on the interval [0,1]. Select $x_0 \varepsilon (a_i, a_{i+1})$ so that $\omega'(x_0) = 1$. This selection is possible by Rolle's theorem. Choose $\phi(x)$ positive for $x \varepsilon [x_0, \omega(x_0)]$ so that

$$\int_{x_0}^{\omega(x_0)} \frac{dy}{\phi(y)} = 1 . \tag{3.1}$$

For $x \varepsilon (\omega(x_0), \omega^2(x_0)]$, set

$$\phi(x) = \omega'(\omega^{-1}(x))\phi(\omega^{-1}(x)) . \tag{3.2}$$

In general, for $x \varepsilon [\omega^n(x_0), \omega^{n+1}(x_0)]$, $n = 1,2,3,\ldots$ set

$$\begin{aligned}\phi(x) &= \omega'(\omega^{-1}(x))\phi(\omega^{-1}(x)) \\ &= \omega'(\omega^{-1}(x))\omega'(\omega^{-2}(x))\ldots\omega'(\omega^{-n}(x))\ \phi(\omega^{-n}(x)) ,\end{aligned} \tag{3.3}$$

where $\omega^{-k}(x) = \omega^{-1}(\omega^{-(k-1)}(x))$ for $k > 1$. For $x \varepsilon (\omega^{-1}(x_0), x_0)$, set

$$\phi(x) = \frac{1}{\omega'(x)} \phi(\omega(x)) .$$

In general, if $x \in [\omega^{-n}(x_0), \omega^{-n+1}(x_0)]$, $n = 0,1,2,\ldots$ set

$$\phi(x) = \frac{\phi(\omega^n(x))}{\omega'(x)\omega'(\omega(x))\ldots\omega'(\omega^{n-1}(x))} . \tag{3.5}$$

Also, define $\phi(a_i) = \phi(a_{i+1}) = 0$. Note that $\lim_{n\to\infty} \omega^{-n}(x) = a_i$ and

$\lim_{n\to\infty} \omega^{n}(x) = a_{i+1}$ for $x \in (a_i, a_{i+1})$.

Lemma 1. ϕ is continuous in [0,1].
The proof is a matter of checking the formulae at the interval end points.

It is well-known (see Kuczma, 1968, p.46) that this $\phi(x)$ is the only continuous solution to (2.1).

Lemma 2.

$$\int_x^{\omega(x)} \frac{dy}{\phi(y)} \equiv 1 \quad . \tag{3.6}$$

Proof. For $x \in (a_i, a_{i+1})$,

$$\frac{d}{dx} \int_x^{\omega(x)} \frac{dy}{\phi(y)} = \frac{\omega'(x)}{\phi(\omega(x))} - \frac{1}{\phi(x)} = 0 \quad .$$

Hence

$$\int_x^{\omega(x)} \frac{dy}{\phi(y)} = \text{constant}$$

Eq.(3.6) then follows from (3.1).

Theorem 2. If $\phi(x)$ is defined as above, then for $x \in (a_i, a_{i+1})$, the initial value problem (1.1) and (1.2) has the unique solution $F(x,t)$ which satisfies $F(x,1) = \omega(x)$.

Proof. Fix $x \in (a_i, a_{i+1})$ and put $y(Z) = F(x,Z)$. By separating variables in (1.1) and integrating with respect to τ, it follows that

$$\int_0^t \frac{1}{\phi(y(\tau))} \frac{dy(\tau)}{d\tau} d\tau = t \quad .$$

Now

$$\int_0^t \frac{1}{\phi(y(\tau))} \frac{dy(\tau)}{d\tau} d\tau = \int_x^{F(x,t)} \frac{dy}{\phi(y)} \quad .$$

Put $t = 1$. Then

$$1 = \int_x^{F(x,1)} \frac{dy}{\phi(y)} \quad ,$$

From Lemma 2,

$$1 = \int_{x}^{\omega(x)} \frac{dy}{\phi(y)} \quad ,$$

so that, since $\phi(y) > 0$, comparing (3.6) and (3.7), we have

$$F(x,1) = \omega(x) \quad .$$

4. EXAMPLE WITH $\omega(x) = \sqrt{x}$

It is useful to present an example illustrating the solution to the functional equation (2.1). Select $\omega(x) = \sqrt{x}$. Then $a_0 = 0$ and $a_1 = 1$. Eq.(2.1) becomes

$$\phi(\sqrt{x}) = \frac{1}{2\sqrt{x}} \phi(x) \quad .$$

The value x_0 is determined by

$$\omega'(x_0) = 1 \ ,$$

so that $x_0 = 1/4$. Also $\omega(x_0) = 1/2$. Hence, by (3.1),

$$\phi(x) = 1/4 \qquad 1/4 \leqq x < 1/2 \quad .$$

In general, for

$$x \ \varepsilon \ (\omega^n(x_0), \omega^{n+1}(x_0)) \quad ,$$

we have

$$\phi(x) = \frac{1}{2^{n+2}} \frac{1}{x^{2^n-1}} \ , \qquad -\infty < n < \infty \quad ,$$

by (3.3) and (3.5).

REFERENCES

[1] Coddington, E.A., Levinson, N.: Theory of Ordinary Differential Equations, McGraw-Hill, 1955, p.10.

[2] Coifman, R.R., Kuczma, M.: On Asymptotically Regular Solutions of a Linear Functional Equation, Aequationes Mathematicae 2, 332-336 (1969).

[3] Kuczma, M.: Functional Equations in a Single Variable, Polska Akademia Nauk, Monografie Matematyczne (1968).

[4] Kucma, M., Choczewski,and Ger, R.: Iterative Functional Equations, to be published by Cambridge University Press.

[5] McKiernan, M.A.: On the Convergence of Series of Iterates, Publicationes Mathematicae Debrecen, 10, 30-39 (1963).

[6] Palis, J.: Vector fields generate few maps, Bull. Amer. Math. Soc., 80, 503-505 (1974).

[7] Targonski, Gy.:Topics in Iteration Theory, Vandenhoeck and Ruprecht, Göttingen, Zürich, p.89 ff. (1981).

[8] Targonski, Gy.:New Directions and Open Problems in Iteration Theory, Bericht Nr.229 (1984), Mathematisch-Statistische Sektion im Forschungszentrum Graz, Austria, 1984.

[9] Utz, W.R.: The embedding of homoemorphisms in continuous flows, Topology Proceedings 6, 159-177 (1981).

[10] Zdun, M.C.: Differentiable Fractional Iteration, Bull.Acad.Pol.Sci., sér.sci. math.astr.et phys., 25, 643-646 (1977).

[11] Zdun, M.: Continuous and differentiable iteration semigroups, Prace Naukowe Uniwersitetu Śląskiego w Katowicach Nr.308 (1979).

W.A. Beyer and P.J. Channell
Los Alamos National Laboratory
Los Alamos, New Mexiko 87545, USA

ON THE BIFURCATION BETWEEN A CHAOTIC AREA OF T^K AND A CHAOTIC AREA OF T

J.C. CATHALA

The existence of absorptive areas and chaotic areas for a second order recurrence with a non unique inverse

$$x_{n+1} = f(x_n,y_n,\lambda) \ , \qquad y_{n+1} = g(x_n,y_n,\lambda) \tag{1}$$

has been mentioned in recent papers [1-3]. In (1), x,y are real variables, λ a real parameter, f,g, are single -valued functions either continuously differentiable or piecewise differentiable.

When the inverse point mapping T^{-1} of T, defined by (1), is a non unique one, the above recurrence admits in general for a range of the parameter λ, a bounded stochastic solution located inside of a closed region of the (x_n,y_n) plane, called chaotic area, limited by segments of critical lines.

For an initial condition belonging to the domain of attraction of the chaotic area, the sequence of consequents, obtained by repeating the application of T, enters after some iterations into the chaotic area and inside of this bounded domain, these points have an apparently erratic movement, without any possibility to escape.

Methods for the construction of chaotic areas and absorptive areas are described in [4-6]. Properties of critical lines and absorptives areas are established in [7-9].

Frequently the chaotic area (d) is a cyclic one of order K, K integer $\geq$ 2. The chaotic area which is an attractive limit set made up of K closed regions without connexion is then given by

$$(d) = \bigcup_{i=1}^{i=K} (d_i) \ . \tag{2}$$

The iterate of every point of the region (d_i), $1 \leq i \leq K$, obtained from (1), is located into (d_{i+1}) which is the consequent of (d_i) under the application of T: $(d_{i+1}) = T(d_i)$. Each of the domains (d_i), constituting the chaotic area is invariant under the application of T^K:$T^K(d_i) = (d_i)$.

The domain of attraction of (d), constituted by the set of points which tend towards the chaotic area is given by

$$(D) = \bigcup_{i=1}^{i=K} (D_i) \tag{3}$$

(D_i) being the domain of attraction of (d_i).

Under the influence of a variation of the parameter λ, the cyclic chaotic area ceases to exist and becomes a chaotic area invariant under the application of T and constituted by a connected domain.

This paper is devoted to the determination of the bifurcation between a cyclic chaotic area of order K and a chaotic area of T. This bifurcation value λ_c of the parameter λ is obtained when a segment of a critical line belonging to the boundary of domains (d_i) constituting (d) has a first-order contact with a segment of invariant curve belonging to the boundary of the domain of attraction (D_i) of these zones if the functions f,g are continuously differentiable (the first-order contact becomes a zero-order contact if one of the two functions f,g or the two functions are piecewise differentiable).

1. LINEAR PIECEWISE RECURRENCE

Consider the linear piecewise recurrence

$$\begin{aligned} x_{n+1} &= y_n \\ y_{n+1} &= y_n - \lambda x_n \quad \text{if} \quad |x_n| \leq 1 \\ y_{n+1} &= y_n + 2x_n - (2+\lambda) \quad \text{if} \quad x_n \geq 1 \\ y_{n+1} &= y_n + 2x_n + (2+\lambda) \quad \text{if} \quad x_n \leq -1 \end{aligned} \tag{4}$$

with $\lambda > 0$.

The recurrence (4) admits three distincts fixed points. The first point $O(x_n = y_n = 0)$ is a stable node if $\lambda < 0.25$, a stable focus if $0.25 < \lambda < 1$, or an unstable focus if $\lambda > 1$. The second and the third fixed points $A(x_n = y_n = 1 + 0.5\lambda)$, $B(x_n = y_n = -(1+0.5\lambda))$ are degenerate saddles with eigenvalues $S_1 = -1$ and $S_2 = 2$; the eigen directions have slopes $p_1 = -1$ and $p_2 = 2$. The straight lines $y_n = -x_n+2 + \lambda(D_1)$ and $y_n = -x_n - 2 - \lambda(D_2)$ are lines of unstable cycles of order two. For $\lambda > 1$, all the cycles of T^K are unstable. The complex bifurcation which occurs for $\lambda = 1$ gives rise to a cyclic chaotic area of order six, constituted by six regions without connexion.

For $1 < \lambda < \lambda_c$, $\lambda_c \simeq 1.12491$, these six regions are unconnected; for $\lambda_c < \lambda \leq 2$, these six regions join, forming a ring shaped attractive zone, constituting the chaotic area. Finally for $\lambda > 2$, the chaotic area disappears becoming a repulsive area [10].

The critical lines of the point mapping T are the two straight lines (LC) and (LC') whose respective equations are $y_n = x_n - \lambda$ and $y_n = x_n + \lambda$. These critical lines will be designated by L and L' in the various figures. These curves are the respective consequents of (LC_{-1}) ($x_n = 1$) and (LC'_{-1}) ($x_n = -1$) (L_{-1} and L'_{-1} in the figures). In the region located between (LC) and (LC') a point has three antecedents of first rank. Outside of this region a point has only one antecedent.

The critical lines of T^k, k integer, are also made up of two branches (LC_i), (LC'_i) with $0 \leq i \leq k-1$, setting (LC_0) =(LC) and(LC'_0) = (LC'). These curves will be designated by L_i and L'_i in the various figures.

Let a_0 and a'_0 be the respective intersections of (LC_{-1}) and (LC) and of (LC'_{-1}) and (LC'). The intersections between (LC_{-1}) and (LC'),(LC'_{-1})and(LC)are designated respectively by M and N. The coordinates of these points are $a_0(x_n = 1, y_n = 1 - \lambda)$, $a'_0(x_n = -1, y_n = -1 + \lambda)$, $M(x_n = 1, y_n = 1 + \lambda)$ $N(x_n = -1, y_n = -1 - \lambda)$.

1.1. Determination of the boundary (F) of (D)

The domain (D) which is the domain of attraction of the chaotic area (d) is constituted by the set of points tending towards the chaotic area. The boundary (F) of (D) which is invariant under the application of T and T^{-1} is defined from the invariant curve crossing the saddles A and B. The segments MC of (D_1) ($x_C = y_M$, $y_C = x_M$) and ND of (D_2) ($x_D = y_N$, $y_D = x_N$) belong to (F) because they are invariant by T and T^{-1}. The boundary (F) is then obtained by taking the antecedents of all rank of these two segments. For $\lambda < 2$, the point M admits two antecedents $M_{-1}(x_n = -1, y_n = 1)$ and $M'_{-1} \equiv C$. $C_{-1}(x_n = -1 - \lambda, y_n = 1 + \lambda)$ and $C'_{-1} \equiv M$ designate the two antecedents of first rank of C. The segment MC possesses, in addition to itself, two antecedents M_{-1} C_{-1} and M_{-1} M. The antecedents of first rank of M_{-1} C_{-1} and M_{-1} M are respectively M_{-2} C_{-2} and M_{-2} C; $M_{-2}(x_n = 2 + 0.5\lambda, y_n = -1)$ and $C_{-2}(x_n = 2 + 1.5\lambda, y_n = -1 - \lambda)$ designating the antecedents of first rank of M_{-1} and C_{-1}.

The boundary (F) is then constituted by the segments MC, M_{-1} M, M_{-i} C_{-i}, M_{-i-1} C_{-i+1}, i = 1,2,.... with $C_0 \equiv C$ and finished by symmetry (Fig. 1).

1.2. Determination of the cyclic chaotic area of order six.

The cyclic chaotic area of order six created by the complex bifurcation of the fixed point O of (e) is constituted, for $1 < \lambda < \lambda_c$, by six regions without connexion. The iterate of every point of the region (d_i), $1 \leq i \leq 6$, obtained from (4), is located in (d_{i+1}). Each of the domains belonging to the chaotic are is invariant under the application of the point mapping T^6. The sequence of the consequents obtained

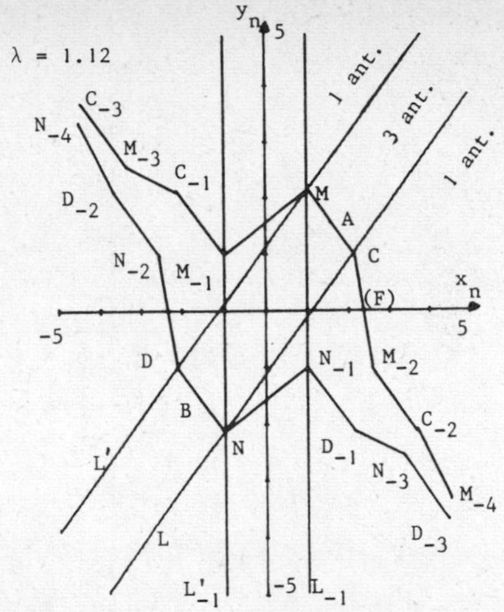

Fig. 1

from (4) and the chaotic area are respectively drawn in Figures 2a and 2b for $\lambda = 1.12$. The Figures 2c and 2d show the enlargement of one of the regions which constitute the chaotic area.

For $\lambda = \lambda_c$, $\lambda_c \simeq 1.12491$, segments of critical lines belonging to the boundary of domains (d_i) have zero-order contact with segments of invariant curves belonging to the boundary of the domains of attraction of these zones.

1.3. Construction of the chaotic area of T.

For $\lambda > \lambda_c$, the six regions join forming a ring shaped attractive zone, constituting a chaotic area (d) which is invariant under the application of the point mapping T: T(d) = (d). This chaotic area (d) has for $\lambda = 1.13$ an external boundary concuring with the boundary of an absorptive area (d'), verifying T(d') = (d'), constituted by the segments b_1a_1 of (LC), a_1a_2 of (LC_1), a_2b_1' of (LC_2), $b_1'a_1'$ of (LC'), $a_1'a_2'$ of (LC_1'), $a_2'b_1$ of (LC_2') (Fig. 3b), with $a_i = T^i(a_o)$, $a_i' = T^i(a_o')$, $b_1 = T(b_o)$, $b_1' = T(b_o')$; b_o and b_o' being the respective intersections between (LC_{-1}) and $a_1'a_2'$ of (LC_1') and between (LC_{-1}') and a_1a_2 of (LC_1). The interior boundary of the chaotic area (d) defines a hole surrounding the repulsive fixed point O. The consequents of a point of this hole excepting only the fixed point, enters after some iterations into (d).

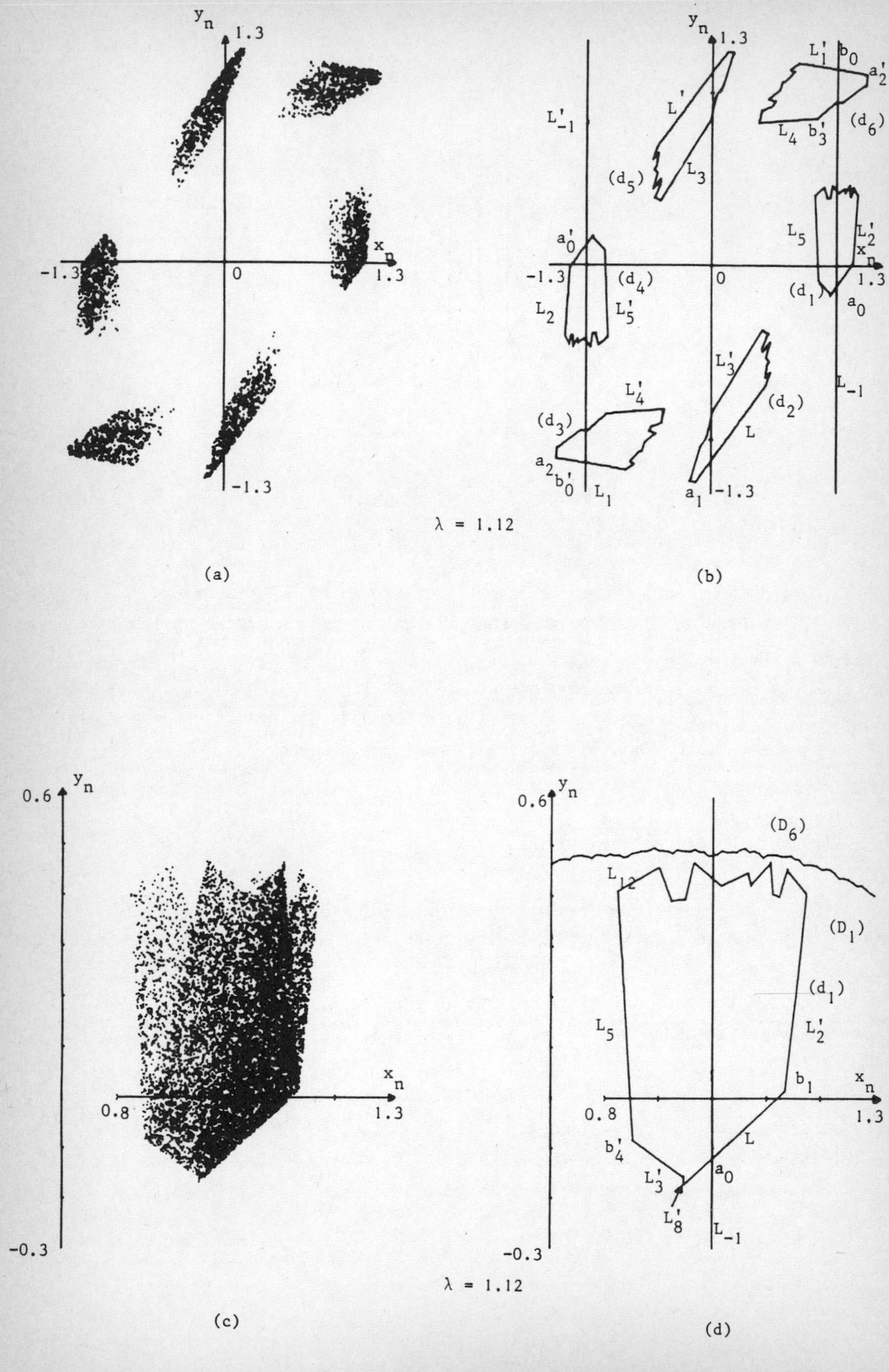

Fig. 2

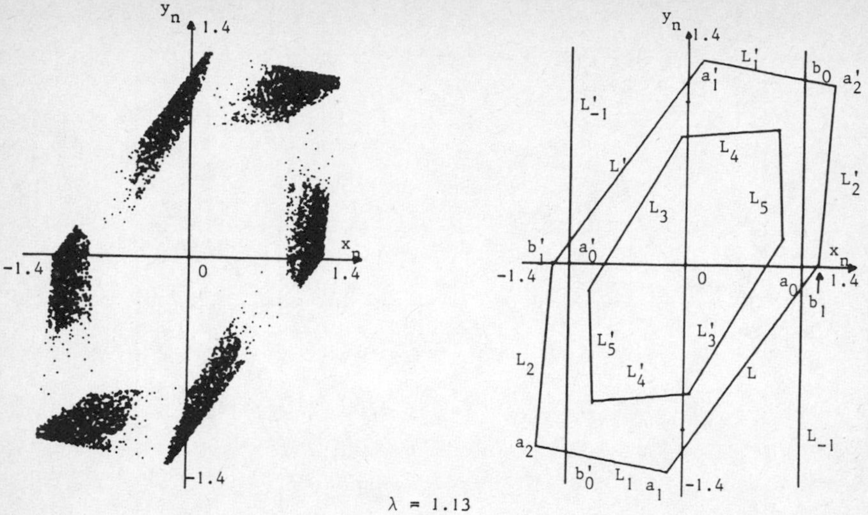

λ = 1.13

(a) (b)

Fig. 3

2. CONTINUOUS RECCURRENCE

Consider the recurrence [11]:

$$x_{n+1} = (1 - \lambda)x_n + y_n \qquad (5)$$
$$y_{n+1} = -0.2\lambda x_n + y_n - \lambda x_n^3 \quad .$$

The fixed point $O(x_n=y_n=0)$ of (5) is a stable node if $\lambda<4/1.8$ and a saddle for $\lambda>4/1.8$. For $\lambda=4/1.8$, one of the eigenvalues of the fixed point O bifurcates and becomes equal to -1. The cycle of order two, created by this bifurcation, bifurcates and gives rise to a cyclic chaotic area of order two which exists for $\lambda < \lambda_c$, $\lambda_c \simeq 2.7845$. For $\lambda > \lambda_c$, the two regions constituting the chaotic area join forming an unique chaotic area.

The critical lines of (5), (LC) and (LC') are given respectively by $y_n = x_n - 2A(1 - 0.8\lambda)/3$ and $y_n = x_n + 2A(1 - 0.8\lambda)/3$, with $A = ((0.8 \;\; - 1)/3\lambda)^{1/2}$. These curves

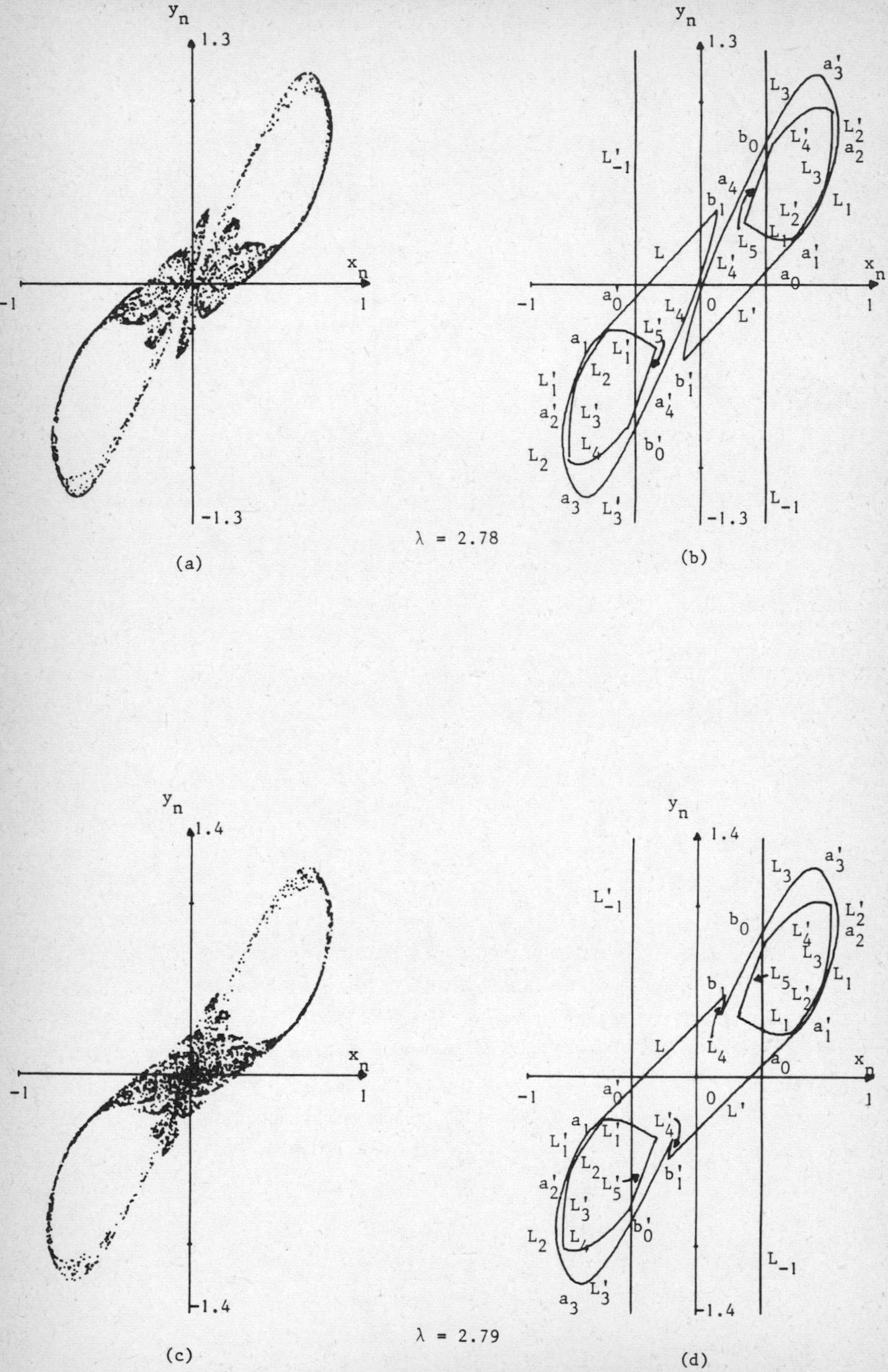

y_n
1.3
x_n
-1
1
-1.3
(a)
λ = 2.78
y_n
1.3
a'_3
L_3
L'_2
b_0
L'_4
a_2
L'_{-1}
L_3
a_4
L_1
b_1
L'_2
L_5
L_1
a'_1
L
L'_4
x_n
a_0
-1
a'_0
L_4
0
L'
1
L'_5
a_1
L'_1
L_2
L'_1
b'_1
a'_2
a'_4
L'_3
L_4
b'_0
L_2
a_3
L'_3
-1.3
L_{-1}
(b)
y_n
1.4
x_n
-1
1
-1.4
(c)
λ = 2.79
y_n
1.4
a'_3
L_3
L'_{-1}
L'_2
b_0
L'_4
a_2
L_3
L_5
L_1
b_1
L'_2
L_1
a'_1
L
L_4
a_0
x_n
-1
a'_0
0
L'
1
a_1
L'_1
L'_4
L'_1
L_2
L'_5
b'_1
a'_2
L'_3
L_4
b'_0
L_2
L_{-1}
L'_3
a_3
-1.4
(d)

Fig. 4

are the respective consequents of (LC_{-1}) $(x_n = A)$ and (LC'_{-1}) $(x_n = -A)$. The region of the phase plane located between (LC) and (LC') is such that a point belonging to this zone possesses three antecedents of first rank. Outside of this region a point has only one antecedent.

Let a_o and a'_o be the respective intersection between (LC_{-1}) and (LC') and between (LC'_{-1}) and (LC).

The chaos being filiform and the density of the points being small, the chaotic area cannot be determinated easily. However a cyclic absorptive area of order two (d'), verifying $(d) \subset (d')$, can be established.

For $\lambda < \lambda_c$ the absorptive area (d') is given by

$$(d') = \bigcup_{i=1}^{i=2} (d'_i) \tag{6}$$

with $T^2(d'_i) \subset (d'_i)$ and $(d_i) \subset (d'_i)$ for $i = 1, 2$.

The sequence of the iterates obtained from (5) and the absorptive area are shown in Figures 4a and 4b for $\lambda = 2.78$.

For $\lambda > \lambda_c$, the cyclic absorptive area of order two ceases to exist and becomes an absorptive area of T, verifying $T(d') \subset (d')$, with $(d) \subset (d')$. The Figure 4c and 4d show the sequence of iterates and the absorptive area for $\lambda = 2.79$.

For $\lambda = \lambda_c$, segments of critical lines belonging to the boundary of domains (d'_i) have first-order contact with segments of invariant curves belonging to the boundary of (D'_i) which designate the domain of attraction of (d'_i). These domains are drawn in Figure 5 for $\lambda = 2.78$.

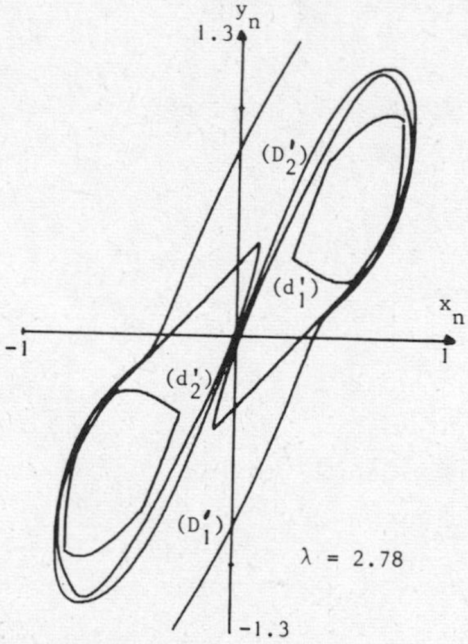

Fig. 5

REFERENCES

[1] Gumowski, I., Mira C.: Solutions chaotiques bornées d'une récurrence, ou transformation poncutelle du 2ème ordre à inverse non unique, C.R. Acad. Sci., Paris, t.285, série A, pp.477-480 (1977).

[2] Gumowski, I., Mira, C.: Dynamique Chaotique, Transformation Ponctuelle, Transition ordre-désordre, Editions Cépadues, Toulouse, (March 1980).

[3] Mira, C.: Complex dynamics in two-dimensional endormophisms, Nonlinear Analysis, vol.4, n°6, pp.1167-1187 (1980).

[4] Barugola, A.: Détermination de la frontière d'une "zone absorbante" relative à une récurrence du deuxième ordre, à inverse non unique, C.R. Acad. Sci., Paris, t.290, série B, pp.257-260 (1980).

[5] Cathala, J.C.: Détermination de zones absorbantes et chaotiques pour un endomorphisme d'ordre deux, Actes du Colloque sur la théorie de l'Itération et ses applications (Toulouse, 17-22 Mai 1982), pp.91-98, Editions du C.N.R.S., Paris (1982).

[6] Cathala, J.C.: Absorptive area and chaotic area in two-dimensional endomorphisms, Nonlinear Analysis, vol.7, n°2, pp.147-160 (1983).

[7] Barugola, A.: Lignes critiques et zones absorbantes pour une récurrence du second ordre à inverse non unique, Actes du Colloque sur la théorie de l'Itération et ses applications (Toulouse, 17-22 Mai 1982), pp.83-89, Editions du C.N.R.S., Paris (1982).

[8] Barugola, A.: Quelques propriétés des lignes critiques d'une récurrence du second ordre à inverse non unique. Détermination d'une zone absorbante, RAIRO, Numérical Analysis, vol.18, n°2, pp.137-151 (1984).

[9] Barugola, A.: On some properties of an absorptive area and a chaotic area for an R^2-Endomorphism, Communication to this Symposium.

[10] Gumowski, I, Mira, C.: Bifurcation déstabilisant une solution chaotique d'un endomorphisme du deuxième ordre, C.R.Acad. Sci., Paris, t.286, série A, pp.427-430 (1978).

[11] Bernussou, J.: Point mapping stability, Pergamon Press (1977).

J. C. Cathala

Université de Provence

Département d'Automatique

et de Dynamique non linéaire

F-13397 Marseille Cedex 13

ON THE DEFINITIONS OF ATTRACTORS

M. Cosnard and J. Demongeot *)

ABSTRACT. We introduce the notion of attractor and present its historical evolution. Then we show that previous definitions are too stringent. We present two equivalent definitions of attractors, show that in this case strange attractors are indeed attractors and give some properties.

Various authors have worked on the definition of the concept of attraction which is of basic importance for a deterministic dynamical system. Smale [16], Thom [18], and Bathia and Szegö [2] proposed several definitions in which the attractor was such that:

- a neighbourhood of the attractor is contained in the domain of attraction
- it satisfies a condition of minimality
- it is in general closed and invariant.

Recently, many numerical experiments on strange attractors have shown that the preceding characteristics are too stringent. A strange attractor is an attractor which is strange: different from a fixed point, a cycle, a closed curve... However it is a stranger fact that such a strange attractor is not in general an attractor, as defined classically. For such an example we refer to Thibault's paper [17] in these proceedings.

Using Bowen's notion of pseudo orbit [3], Conley [4] and Ruelle [14] proposed a more general approach of the concept of attractor.

In this paper, we introduce the notion of attractor and present its historical evolution. Then we show that previous definitions are too stringent and try to analyse what we would like to call an attractor. We deduce that an attractor must be invariant under the double dual action of taking its basin of attraction and taking the limit set of this basin.

Lastly we present two equivalent definitions of attractors, show that in this case strange attractors are indeed attractors and give some properties.

1. PREVIOUS DEFINITIONS

Let f be a continuous selfmap of a compact metric space (E,d). We call L(x) the set of limit points of $f^n(x)$:

---

*) Presented by M.Cosnard

$$L(x) = \{ y \in E / \exists n_i \to +\infty ; \lim_{i \to +\infty} d(f^{n_i}(x), y) = 0 \} .$$

We recall three different definitions of attractor among others:

Definition 1 (Thom [18]). A is an attractor if

1. almost all trajectories of A are dense in A
2. $\exists \{V_i\}_{i \in I}$ fundamental system of neighbourhood of A such that

   2.1 $\forall \ M \subset V_i, \lim_{n \to \infty} d(f^n(M), A) = 0$

   2.2 $\forall \ M \subset V_i, \lim_{n \to \infty} f^{-n}(M) \cap A \neq 0 \Rightarrow M \subset A.$

Definition 2 (Guckenheimer and Holmes [11]). A is an attractor if

1. there exists a trajectory of A dense in A
2. $\exists \ V(A), \forall \ n \geq 0, f^n(V) \subset V$ and $\forall \ x \in V, L(x) \subset A.$

Definition 3 (Ruelle [14]). A is an attractor (included in an attracting set) if

1. $\exists \ V(A), A = \bigcap_{n \geq 0} f^n(V)$
2. A is an equivalence class of the Ruelle-Bowen equivalence relation.

These three definitions are constructed using the same model: existence of a neighbourhood contained in the domain of attraction, and a condition of minimality. It is not yet known if these definitions are equivalent.

However the first requirement is too stringent: the neighbourhoods of an attractor can contain trajectories which remain in these neighbourhoods but do not converge to the attractor: an example is constructed in part 3.

The condition of minimality is obtained through the use of dense trajectories or Ruelle-Bowen relation. In the following we shall use this relation which is more general than the density of trajectories.

2. RUELLE-BOWEN EQUIVALENCE RELATION

Let x and y belong to E. We say that there exists a pseudo trajectory from x to y and write $x \to y$ if

$$\forall \ \varepsilon > 0, \ \exists \ z_1, \ldots, z_{j(\varepsilon)} \text{ and } \exists \ n_1, \ldots, n_{j(\varepsilon)} \geq 1$$

such that

$$z_1 = x, \ z_j = y, \ d(f^{n_i}(z_i), z_{i+1}) \leq \varepsilon \quad i = 1, \ldots, j-1$$

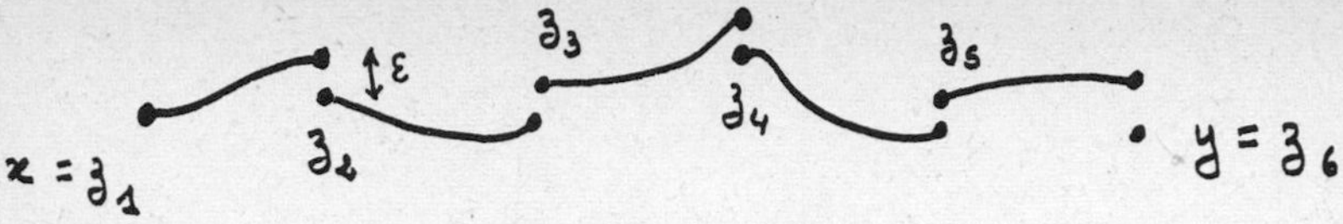

Fig. 1

We call heart of E and write C(E) the set of x such that there exists a pseudo trajectory from x to x:

$$C(E) = \{x \in E / x \to x\}.$$

Consider the following relation R on C(E):

$$\forall\ x, y \in C(E),\ x\, R\, y \Leftrightarrow\ x \to y \text{ and } y \to x \quad .$$

It is an equivalence relation, called Ruelle-Bowen equivalence relation.

3. A CANTOR ATTRACTOR

The Feigenbaum functional equation [9] is the following: find a continuous unimodal selfmap f of [-1,+1] such that

$$f(x) = (-1/\alpha) f^2(-\alpha x);\ \alpha = -f(1) > 0 .$$

This equation is close to that studied by Dubuc [8] in these proceedings. In Cosnard [5], an algorithm for constructing solutions of this equation is proposed and the iterative behaviour of such solutions is studied.

These functions (an example is presented in Fig. 2) admit a cycle of order 2^i for all integer i and a Cantor set which can attract almost all trajectories in the sense of Lebesgue measure. However given a neighbourhood of this Cantor set, there exists i such that the cycles of order greater than 2^i are contained in this neighbourhood.

This is an example of an attractor whose basin of attraction is of Lebesgue measure one but does not contain any neighbourhood of the attractor.

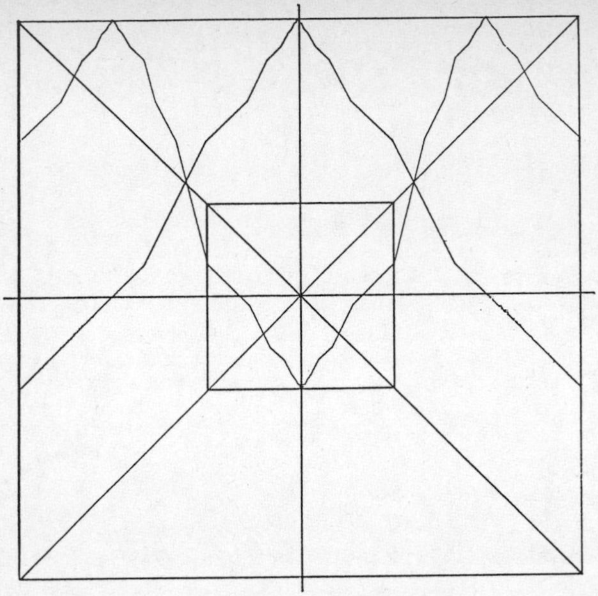

Fig. 2

4. INTUITIVE DEFINITION AND EXAMPLES

We would like that an attractor remains invariant under the double dual operations of:

- taking its basin of attraction
- taking the limit set of this basin.

Moreover it must satisfy a condition of maximality with respect to the connexity under the iteration and a condition of minimality in order that the union of two attractors is not an attractor.

We would like to stress that in all general definitions there is some kind of personal touch. For examples:

- in Figure 3.1, is the attractor the whole grid or does there exist four attractors?
- in Figures 3.2 and 3.3 is the attractor the whole circle or each fixed point?
- in Figure 3.4 the whole grid with or without the bottom line or each horizontal line?

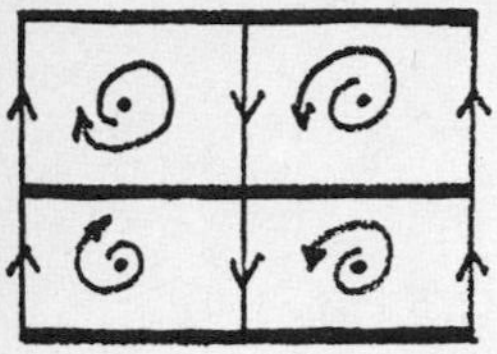

3.1

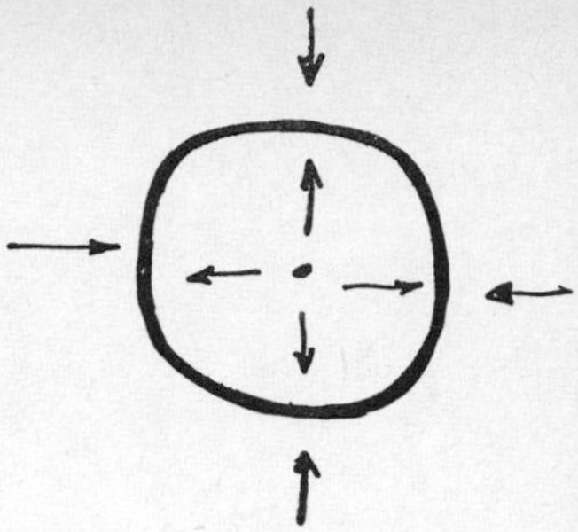

3.2

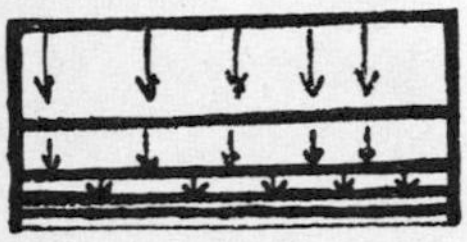

3.4

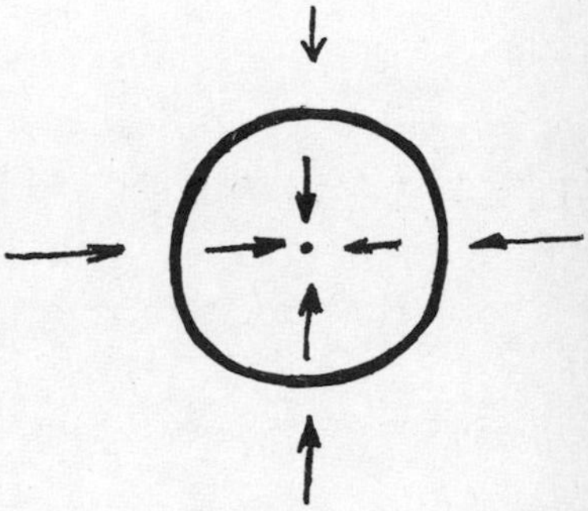

3.3

Fig.3. The thick lines correspond to fixed points

Moreover example 3.3 sets up the case of semi stable attractors and example 3.4 the case of non compact attractors.

We decided to choose arbitrarily the whole circle and the whole grid in each case by the use of a maximality condition.

5. FIRST DEFINITION

If $A \subset E$, we define $L(A) = \bigcup_{x \in A} L(x)$. The basin of attraction of A is defined to be

$$B(A) = \{x \in E \setminus A / L(x) \subset A\}$$

x of E is called super non recurrent if

$$\nexists\, y \in E; \quad x \in L(y) \quad .$$

Let S be the set of super non recurrent points. We write $B_S(A) = S \cap B(A)$. L, B, and B_S are considered as operators on $\mathcal{P}(E)$.

Properties

1. $f(S) = S$; $S = E - L(E)$.
2. $(L \circ B)^2 = L \circ B$; $(L \circ B_S)^2 = L \circ B_S$.
3. $\forall x \in E, \forall y, \quad z \in L(x); \quad y \rightarrow z$. We say that L(x) is chain recurrent.
4. $L(E) \subset C(E)$.

Let C_i, $i \in I$, be the equivalence classes of C(E)/R. Hence L(x) is chain recurrent if and only if

$$\exists\, i \in I; \qquad L(x) \subset C_i \quad .$$

Definition I. $A \neq \emptyset \subset E$ is a weak attractor if:

1. $L \circ B_S(A) = A$.
2. $\nexists A'$, chain recurrent such that A' contains strictly a chain recurrent component of A and $L \circ B_S(A \cup A') = A \cup A'$.
3. $\nexists A''$ such that $A'' \subset A$ and A" satisfies 1. and 2.

Lemma 1. If A is a weak attractor then

1. $f(A) = A$.
2. $L(A) \subset A$.
3. A is chain recurrent.

As an illustration, let us consider the dynamical system of Fig. 4:

$$S = \{ (x,y) \in E / x \neq 0 \}; \quad A = \{ (x,y) \in E / x = 0 \} \quad .$$

A is a weak attractor but (0,0) is not an attractor.

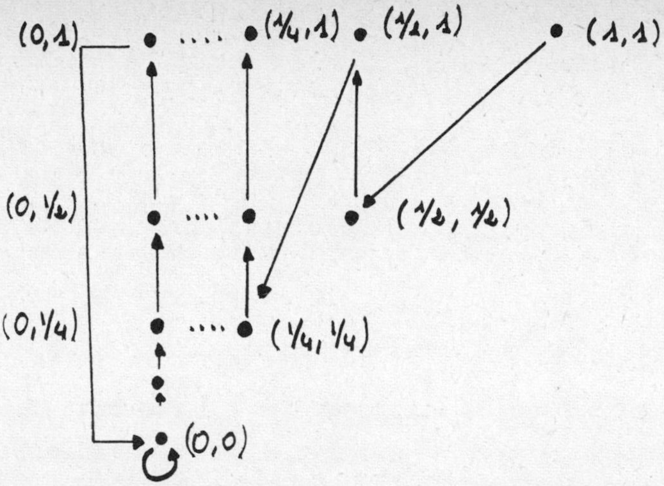

Fig. 4

6. SECOND DEFINITION

We call attracting set, the set A defined by

$$\mathcal{A} = \bigcup_{A = L \circ B(A)} A .$$

We call super attracting set, the set $\mathcal{A}_S$ defined by

$$\mathcal{A}_S = \bigcup_{A = L \circ B_S(A)} A .$$

Lemma 2.

1. $\mathcal{A} = \bigcup_{A = L \circ B(A)} A = \bigcup_{A \in E} L \circ B(A) = L(N)$.

2. $\mathcal{A}_S = \bigcup_{A = L \circ B_S} A = \bigcup_{A \in E} L \circ B_S(A) = L(S) = L(E-L(E))$.

3. $A_S \subset A \subset L(E)$.

4. $A = L_o\, B(A)$ and $A_S = L_o\, B_S(A_S)$.

A_S can be strictly included in A as is shown in an example of Cosnard and Demongeot [6].

Definition II. $A = \emptyset \subset E$ is a weak attractor if it is a chain recurrent component of A_S.

Equivalence theorem: Definition I and II are equivalent.

Definition III. $A \neq 0 \subset E$ is a strong attractor if A is a weak attractor such that $B(A) \neq 0$.

A is a super strong attractor if $A \cup B(A)$ contains a neighborhood of A.

7. PROPERTIES OF ATTRACTORS

P1.: A necessary and sufficient condition of existence of a weak attractor is that $S \neq 0$, i.e., $E \neq L(E)$.

P2.: If A is an attractor in the sense of definition 1, 2 or 3, then it is a super strong attractor.

P3: The CANTOR attractor described in section 3 is a strong attractor.

REFERENCES

[1] Birkhoff, G.D.: Dynamical systems, AMS Colloqu. Pub. 9, New York (1927).

[2] Bhatia, N.P., Szegö, G.P.: Dynamical systems: stability theory and applications, Lect. Notes Math. 35, Springer Verlag (1867).

[3] Bowen, R.: On Axiom A diffeomorphisms,Reg.Conf. Series Math.35, AMS Providence (1878).

[4] Conley, C.: Isolated invariant sets and the Morse index, Reg.Conf.Series Math. 38, AMS Providence (1978).

[5] Cosnard, M.: "Etude des solutions de l'équation fonctionnelle de Feigenbaum", Actes Coll. Dijon Astérisque 98-99, p.143-152 (1983).

[6] Cosnard, M., Demongeot, J.: "Attracteurs: une approche déterministe", C.R.Acad.Sc.Paris 300, 15 (1985) p.551-555.

[7] Demongeot, J.: Systèmes dynamiques et champs aléatoires application en biologie fundamentale, thèse, Grenoble (1983).

[8] Dubuc, S.: "Une équation fonctionelle pour diverses constructions", these proceedings.

[9] Feigenbaum, M.J.: "The universal metric properties of nonlinear transformations", J. Stat. Phys. 21, 6,p.669-709 (1979).

[10] Garrido, L., Simo C.: "Some ideas about strange attractors", Lect. Notes Phys. 179, Springer Verlag (1983).

[11] Guckenheimer, J., Holmes P.: Nonlinear oscillations, dynamical systems and bifurcations of vector fields, Applied Math. Science 42, Springer Verlag (1983).

[12] Lozi, R.: Modèles mathématiques qualitatifs simples et consistants pour l'étude de quelques systèmes dynamiques expérimentaux, thèse, Nice (1983).

[13] Nemytskii, V.V., Stepanov, V.V.: Qualitative theory of differential equations, Princeton Univ. Press, New York (1960).

[14] Ruelle, D.: "Small random perturbations and the definition of attractors", Comm. Math. Phys. 82, p.137-151 (1981).

[15] Sinai, Y.G.: "The stochasticity of dynamical systems", Sel. Math. Sov. 1, p.100-119 (1981).

[16] Smale, S.: "Differentiable dynamical systems", Bull. AMS Soc. 73, p.747-817 (1967).

[17] Thibault, R.: "Competition of a strange attractor with attractive cycle", these proceedings.

[18] Thom, R.: Modèles mathématiques de la morphogénèse, C.Bourgeois Ed., Paris (1980).

[19] Williams, R.F.: "Expanding attractors", Publ. Math. IHES 43, p.169-203 (1974).

M. Cosnard and J. Demongeot
Laboratoire TIM3
Université de Grenoble BP 68
F-38402 Saint Martin d'Heres

FUNCTIONAL EQUATIONS CONNECTED WITH PECULIAR CURVES

Serge Dubuc

Our concern is the functional equation

$$x(t) = f(t,x(b(t))) \tag{1}$$

where t belongs to T, x belongs to X, f(t,x) is a bivariate function with values in X and b(t) is a transformation of T, the unknown function is the function x(t) from X to T. This equation has been considered by many authors, Kuczma [7] is one of those. Our aim is to give many examples where this equation occurs very naturally. Classical non-differentiable functions, as defined by Weierstrass, Knopp, van der Waerden, Hildebrandt, Sierpiński, can be put in this frame. The same is true for the dragon curve of Harter-Heightway, Mandelbrot-von Koch curves and other curves. We will introduce a new family of curves whose main property is that they are partially self-similar. At the end of this article, we will describe an iterative scheme for the graphical computation of the curves related to this functional equation.

1. SOME CURVES WITHOUT TANGENT

a) Weierstrass's function [12]. The function W(t) is given by the series $\sum_{n=0}^{\infty} k^n \cos(\omega^n t)$ where $k \in]0,1[$ and $\omega > 0$. This function is the unique bounded solution to the equation (1) where T and X are the real axis, $b(t) = \omega t$ and $f(t,x) = \cos t + kx$.

b) Cellērier's function [1]. C(t) is the series $\sum_{n=1}^{\infty} a^{-n} \sin (a^n t)$. C(t) satisfies the equation (1) where T and X are the real axis, b(t) = at and $f(t,x) = (\sin at + x)/a$.

Figure 1 is the mixture of Weierstrass's function with Cellērier's one: it is the trace of the curve (W(t), sin t + C(t)) as functions parameters are k = 1/3, $\omega = 3$ and a = 3. More generally, the unique bounded solution to the functional equation (1), where T is the real axis, X is the complex plane, $b(t) = \omega t$ and $f(t,x) = e^{it} + kx$, is the function $\sum_{n=0}^{\infty} k^n e^{i\omega^n t}$.

c) Knopp's function [6]. It is the function coming from the series $\sum_{n=0}^{\infty} a^n \psi(b^n t)$ where $\psi(t)$ is the distance of t to the nearest integer. Knopp's function satisfies also the functional equation (1): X and T are the real axis, b(t) = bt and $f(t,x) = \psi(t) + ax$. van der Waerden [10] and Hildebrandt [5] were involved in Knopp's function in the following cases: a = 1/10 and b = 10, then a = 1/2 and b = 2. Figure 2 is the graph of the function of van der Waerden.

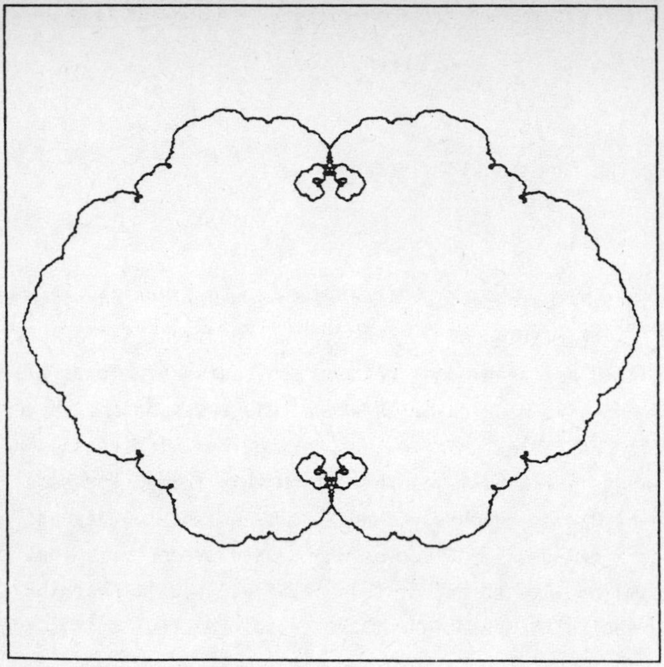

Fig. 1. Mixture of functions of Weierstrass and Cellérier

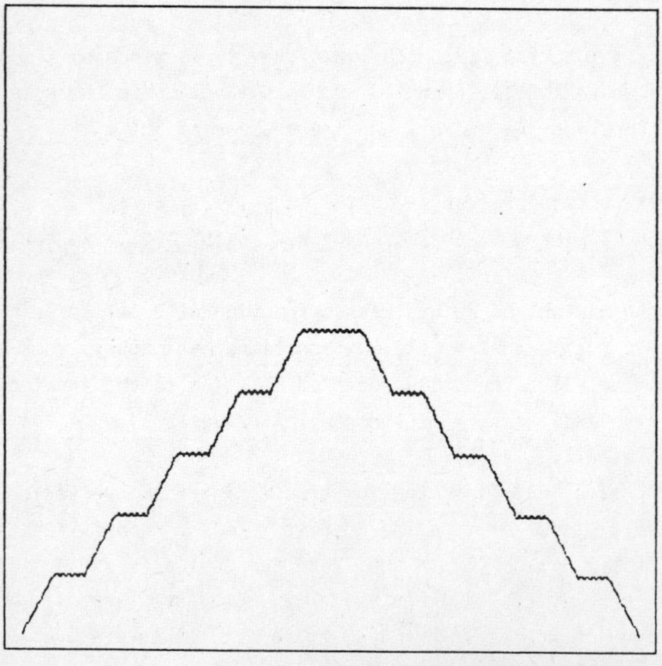

Fig. 2. van der Waerden's function

d) Sierpiński's function [9]. As example of a function with a derivative at no point, Sierpiński defined the function $S(t)$, the unique bounded solution of the equation (1) where T and X are the real axis, $b(t) = 6t+1+(-1)^{[t]}/2$ and $f(t,x) = ((-1)^{[t]}+x)/2$.

e) Dragon curve of Harter-Heightway [4]. One defines two contractions L and R of the plane: $L(x,y) = ((x+y)/2,(-x+y)/2)$ and $R(x,y) = (1-(x-y)/2,-(x+y)/2)$. The dragon curve is defined as the graph of the function $H(t)$, the unique bounded solution of the functional equation (1) where $T = [0,1]$, X is the plane, $b(t) = 2t$ and $f(t,x,y) = L(x,y)$ if $t < 1/2$, $b(t) = 2(1-t)$ and $f(t,x,y) = R(x,y)$ for other values of t.

f) Curves of von Koch-Mandelbrot. Mandelbrot [8] has extended a geometrical construction of von Koch [11]. His suggestion is to construct self-similar plane curves. Such a curve is built with b arcs similar to the entire curve. For b+1 points of the plane X, P_0, $P_1,\ldots,P_b$, he introduces b plane similarity transformations S_i: the image of the segment P_0,P_b by S_i is the segment P_{i-1},P_i. We set $T = [0,1)$, $b(t)$ is the fractional part of bt and $f(t,x,y) = S_i(x,y)$ if $(i-1)/b \leq t < i/b$. Any solution of functional equation (1) is a self-similar curve.

g) Siamese sisters. We bring back a curve already described in [2]. $x(t)$ is the unique bounded solution of the functional equation (1) where X is the real axis, T is $[0,1)$, $b(t)$ is chosen as the fractional part of 4t. $f(t,x)$ is defined as follows:

$$-x/4 \quad \text{if} \quad 0 \leq t < 1/4 \qquad -1/4 + 3x/4 \quad \text{if} \quad 1/4 \leq t < 1/2$$
$$1/2 - x/4 \quad \text{if} \quad 1/2 \leq t < 3/4 \qquad 1/4 + 3x/4 \quad \text{if} \quad 3/4 \leq t < 1 \quad .$$

The parametric curve $(x(t),y(t))$ with $y(t) = 1-x(1-t)$ has been called the curve of Siamese sisters. Figure 3 is the graph of the function $y(t)$, figure 4 is the trace of the parametric curve. In the last figure, the range of the x-axis is $[-1,1]$, and the range of the y-axis is $[0,2]$.

2. PARTIALLY SELF-SIMILAR CURVES

We use a variation of a theme coming from Mandelbrot. The basic property of these curves is that they can be split in arcs, each of them are contractions of other parts of the curve, these parts being union of those arcs.

If T is a finite cellular complex of dimension one without isolated points, T is then made of p compact simple arcs $A_1, A_2,\ldots,A_p$ such that any common point of two of these arcs is their (common) endpoint. $t_1,t_2,\ldots,t_q$ will denote points of T which are endpoints of A_i. With each arc A_i, is given a continuous function b_i from A_i to T such that endpoints of A_i are sent onto $t_{\phi(i)}$ and $t_{\psi(i)}$. The following assumptions hold: X is a arc-connected, complete metric space, $x_1,x_2,\ldots,x_q$ belong to X and S_i are contractions of X such that $S_i(x_{\phi(i)})$ and $S_i(x_{\psi(i)})$ are precisely x_j and x_k if

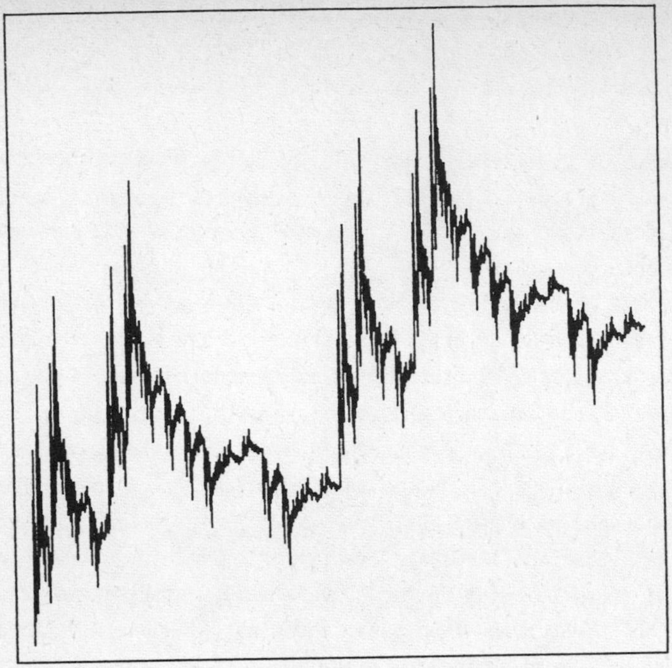

Fig. 3. y-component of Siamese sisters

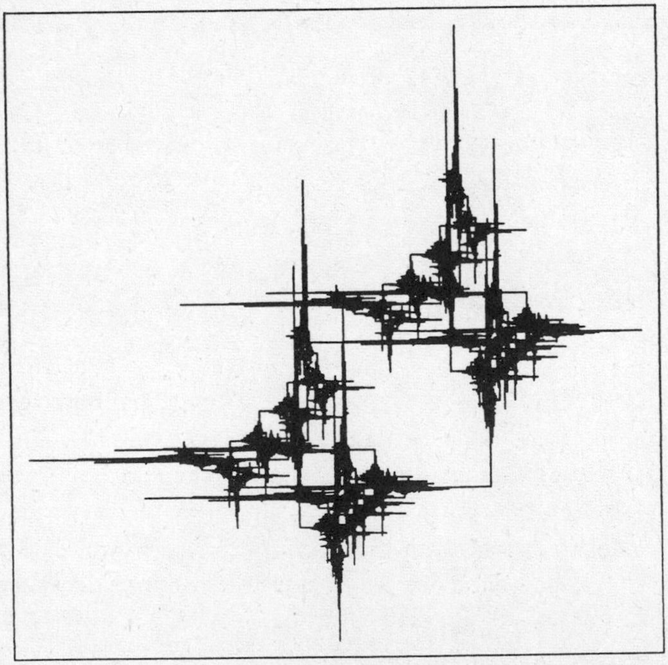

Fig. 4. Siamese sisters

t_j and t_k are endpoints of the arc A_i. Under these assumptions, the following result can be proven (see [3]).

Theorem. There is one and only one bounded function from T to X such that for any integer of [1,p] and for any point t of A_i, $x(t) = S_i(x(b_i(t)))$. This function x(t) is continuous.

It is possible to connect Sierpiński's function or the dragon curve to the last theorem. As another application, we define a new family of curves: partially self-similar closed curves.

This family uses two parameters p and r. The first one, p, is an integral value greater or equal to 4. The parametric space T is [0,1] and is split into p intervals $A_i = [i/p,(i+1)/p]$, $0 \leq i \leq p-1$. Both endpoints of T, 0 and 1, are identified. The functions b_i of the previous theorem come from the second parameter r, a relative integer. If t belongs to A_i, $b_i(t)$ is the fractional part of $i/p + r(t-i)/p$. It is assumed that the remainder of the division of r by p is different of 0, 1 and p-1. The choice of X is the cartesian plane. Points P_i of X related to the values $\{i/p\}_{i=0}^{p-1}$ of T are the vertices of regular polygon with p sides: $P_i = (\cos(2\pi i/p), \sin(2\pi i/p))$. If $j = i$ and $k/p = b_i((i+1)/p)$, the transformation S_i is the similarity which sends the segment $[P_j, P_k]$ onto the segment $[P_i, P_{i+1}]$. Already given restrictions on r ensure that S_i is a contraction. With those specifications, the curve defined in the theorem is a continuous closed curve. Figures 5 and 6 are two cases of these curves: in the first case, parameters are p = 5 and r = -2 and in the second, the parameters are p = 5 and r = -3.

3. NUMERICAL COMPUTATION OF THE SOLUTION OF EQUATION (1)

Two new ideas are helpful in order to find the numerical solution of the functional equation (1). The first idea is to use stronger and more regular contractions. Let us come back to the equation $x(t) = f(t,x(b(t)))$. We introduce by induction a sequence of contractions $f_n(t,x)$: $f_1(t,x) = f(t,x)$ and $f_{n+1}(t,x) = f(t,f_n(b(t),x))$. If $\{P_k\}_{k=1}$ is a partition of the space T, if t belongs to P_k, we set $b^*(t) = b_k(t)$, the k th iterated power of b and we set $f^*(t,x) = f_k(t,x)$. It is obvious that x(t) is also a solution to the functional equation

$$x(t) = f^*(t,x(b^*(t))) \quad . \qquad (1^*)$$

If δ is a positive number and if t belongs to T, we define N(t) as the first integer k for which the contraction factor of the transformation $f_k(t,x)$ of X is smaller than δ. The function N(t) gives a partition of T: $P_k = \{t : N(t) = k\}$. Every function $x \to f^*(t,x)$ is a contraction whose factor is smaller than δ.

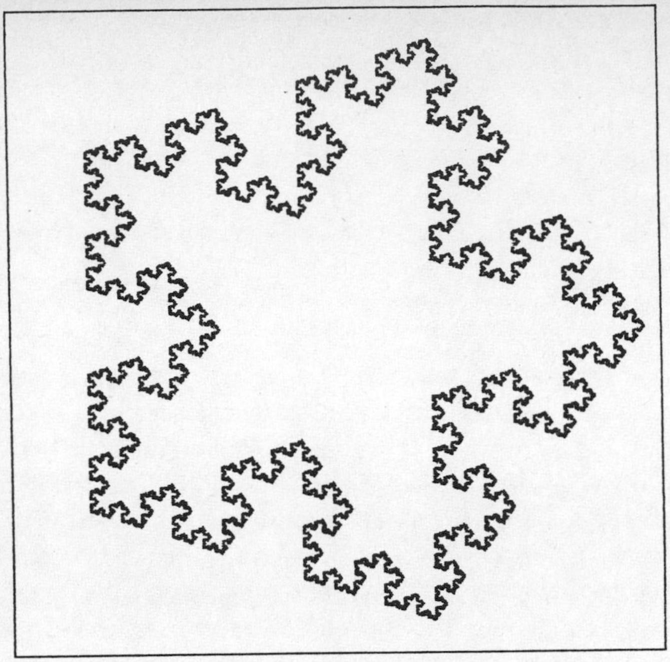

Fig. 5. Binary divisions of a pentagon

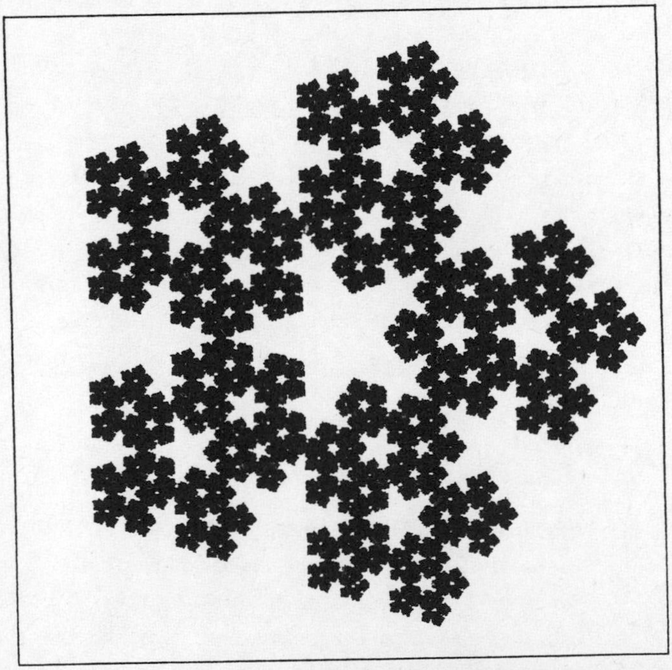

Fig. 6. Ternary divisions of a pentagon

The second idea is simple in itself. We start with an approximate solution $x_0(t)$ of the equation (1*). We set $x_1(t) = f^*(t,x_0(b^*(t)))$ and $x_2(t) = f^*(t,x_1(b^*(t)))$. In most cases, $x_2(t)$ will be a very good approximation of the solution of (1). Indeed, the distance of $x_2(t)$ to this solution will be bounded by the distance of $x_1(t)$ to $x_0(t)$ multipled by $\delta^2/(1-\delta)$. One advantage of this method is to avoid intermediate computations.

In fact, figures 3, 4, 5 and 6 were plot by using these two previous ideas. This article is a summary of [3], with some more details. All the figures other than the fourth one are new. They have been computed with a microcomputer APPLE III, the printer was an EPSON MX-80. High resolution graphics were done within an area of 480 x 864 pixels. In figure 1, there are 3^8 = 6561 evaluations of pixels; after removal of repeated pixels, they were 3664. Figure 2 has been got after 2000 evaluations of van der Waerden's function; it was then possible to find 480 columns of pixels with a mean height of 3 pixels; the number of pixels in the figure 2 is 1456. In figure 3, the function has been computed at 1024 places and 480 columns of pixels with a mean height of 46 pixels were found; the whole figure contains 22135 pixels. Figure 4 is built with 15712 vertical groups of pixels (mean height: 5,7) and 11232 horizontal groups of pixels (mean width: 4,9). Many of these groups interfere. I do not know the precise number of pixels in that figure. Figure 5 comes from the computation of 64 x 320 = 20480 points. Figure 6 is the result of careful choice of 5 x 202 x 285 = 287850 points.

REFERENCES

[1] Cellérier, C.: Note sur les principes fondamentaux de l'analyse. Bull. des Sc. Math. (2), vol. XIV première partie 142-160 (1890).

[2] Dubuc, S.: Une foire de courbes sans tangentes. Actualités mathématiques. Actes VI ème Congrès Mathématiciens d'Expression Latine. Gauthier-Villars. Paris 99-123 (1982).

[3] Dubuc, S.: Une équation fonctionelle pour diverses constructions géométriques. Ann.sc.math.Québec vol.9 no.2 (1985), to appear.

[4] Gardner, M.: Mathematical games. Scientific American 216 (1967), mars 124-125 et avril 118-120.

[5] Hildebrandt, T.H.: A simple continuous function with a finite derivative at no point. Amer.Math.Monthly 40, 547-548 (1933).

[6] Knopp, K.: Ein einfaches Verfahren zur Bildung stetiger nirgends differenzierbarer Funktionen. Math. Zeitschrift 2, 1-26 (1918).

[7] Kuczma, M.: Functional equations in a single variable. Polish Scientific Publishers, Warsaw, 1968.

[8] Mandelbrot, B.B.: The fractal geometry of nature. W.H.Freeman, San Francisco, 1982.

[9] Sierpiński, W.: Sur deux problèmes de la théorie des fonctions non dérivables. Bull.Intern.Acad.Sci.Cracovie A, 162-182 (1914).

[10] van der Waerden, B.L.: Ein einfaches Beispiel einer nichtdifferenzierbaren stetigen Funktion. Math.Zeitschrift 32, 474-475 (1930).

[11] von Koch, H.: Sur une courbe continue sans tangente obtenue par une construction géométrique élémentaire. Arkiv för Matematik, Astronomie och Fysik 1, 681-704 (1904).

[12] Weierstrass, F.: Uber continuirliche Functionen eines reellen Arguments die für keinen Werth des letzteren einen bestimmten Differentialquotienten besitzen. Mathematische Werke II, 71-74 (1872).

Serge Dubuc
Dept.de mathématiques et de statistique
Université de Montréal
C.P. 6128
Montréal, Québec
H3C 3J7 Canada

ITERATION AND ANALYTIC CLASSIFICATION OF LOCAL DIFFEOMORPHISMS OF $\mathbb{C}^{\nu}$

Jean Ecalle

CONTENTS

§1. INTRODUCTION. RESURGENT FUNCTIONS

Two fundamental problems arise in connection with the local diffeomorphisms ("diffeos" for short) of $\mathbb{C}^{\nu}$: how to iterate them (formally or analytically) and how to classify them (again, formally or analytically). Actually, although those twin problems are largely equivalent, the second one is slightly more comprehensive. We shall therefore start with classification (§2 and §3) before tackling iteration (§4).

Everything rests on the use of resurgent functions. Roughly speaking these are divergent power series:

$$\phi(z) = \sum a_n z^{-n} \tag{1.1}$$

whose Borel transforms:

$$\hat{\phi}(\zeta) = \sum a_n \zeta^{n-1}/(n-1)! \tag{1.2}$$

converge in a neighbourhood of the origin and give rise, through analytic continuation in ζ, to analytic functions which, though usually multivalued, never exhibit any cuts (that is to say, functions with isolated but otherwise arbitrary singularities). These functions are an algebra under the formal multiplication of the originals ϕ or, what amounts to the same, under the convolution $*$ of the Borel transforms:

$$(\hat{\phi} * \hat{\psi})(\zeta) = \int_0^{\zeta} \hat{\phi}(\zeta_1)\hat{\psi}(\zeta-\zeta_1)d\zeta_1 \qquad (\zeta \sim 0) \tag{1.3}$$

On this algebra one introduces a whole array of linear operators Δ_ω which in some sense measure the singular behaviour of each $\hat{\phi}$ at its singular points ω and which are so defined as to be derivations:

$$\Delta_\omega(\hat{\phi} * \hat{\psi}) = (\Delta_\omega\hat{\phi}) * \hat{\psi} + \hat{\phi} * (\Delta_\omega\hat{\psi}) \quad . \tag{1.4}$$

Or equivalently, keeping the same symbols Δ_ω but reverting to the originals ϕ:

$$\Delta_\omega(\phi.\psi) = (\Delta_\omega\phi).\psi + \phi.(\Delta_\omega\psi) \quad . \tag{1.5}$$

The operators Δ_ω are known as <u>alien derivations.</u> They are subject to no a priori commutation constraints.

All these constructions extend to power series far more general than (1.1). For precise statements see [E1].

§2. DIFFEOMORPHISMS TANGENT TO THE IDENTITY. THEIR FORMAL CLASSIFICATION

To illustrate the method, we shall deal with the case of local diffeos of $\mathbb{C}^\nu$ that are tangent of order $p(p \geq 1)$ to the identity, hence of the form:

$$f : x_i \to f_i(x) = x_i + o(x^p) \qquad (1 \leq i \leq \nu) \quad . \tag{2.1}$$

More precisely, those diffeos may be written:

$$f : x_i \to f_i(x) = x_i + Q_i(x) + S_i(x) \qquad (Q_i \in \mathbb{C}[x],\ S_i \in \mathbb{C}\{x\}) \tag{2.2}$$

the $Q_i(x)$ being homogeneous polynomials of degree p+1 in $x_1,\dots,x_\nu$ and the $S_i(x)$ being functions holomorphic near 0 and vanishing of order p+1 at 0. The integer p is said to be the <u>level</u> of the diffeo f.

The rational self-mapping

$$Q : \gamma = (\gamma_1,\dots,\gamma_\nu) \to Q(\gamma) = (Q_1(\gamma),\dots,Q_\nu(\gamma)) \tag{2.3}$$

ofthe projective space $\mathbb{P}_{\nu-1}(\mathbb{C})$ is known as <u>directive mapping</u> of f. Its fixed points are called <u>eigenradii</u> of f. For each eigenradius γ one denotes:

$$1 - \lambda_1(\gamma),\ 1 - \lambda_2(\gamma),\dots,\ 1 - \lambda_{\nu-1}(\gamma) \tag{2.4}$$

the ν-1 eigenvalues of the Jacobian of Q at γ. The scalars λ_j are called directors of the eigenvalues.

To avoid extraneous complications and get easily enuntiable results, let us make two (non essential and largely expendable) genericity assumptions:

(A_1) Each fixed point (eigenradius) of the directive mapping Q is simple.

(A_2) The directors $\lambda_1,\dots, \lambda_{\nu-1}$ associated with every eigenradius γ are irrational and mutually independent over $\mathbb{Z}$.

The following statement disposes of the formal classification:

Proposition 1: (Formal invariants)

For each eigenradius γ there exists a formal change of variables

$$x_i = h_i(y_0,y_1,\dots,y_{\nu-1}) \qquad (1 \leq i \leq \nu;\ h_i(y) \in \mathbb{C}[[y]]) \tag{2.5}$$

that is invertible, unique up to dilatations:

$$y_i \to c_i\ y_i \qquad (0 \leq i \leq \nu-1;\ c_i \in \mathbb{C}) \tag{2.6}$$

and such that

$$f = \exp X \tag{2.7}$$

where

$$p(1+\rho y_0^p)X = [-y_0^{p+1}+R_0(y)]\frac{\partial}{\partial y_0} + \sum_{i=1}^{\nu-1} [(\lambda_i-1)y_0^p y_i + R_i(y)]\frac{\partial}{\partial y_i}$$

$$R_0,R_1,\dots,R_{\nu-1} \in \mathbb{C}[[y_0,y_1,\dots,y_{\nu-1}]] \tag{2.8}$$

with a scalar $\rho = \rho(\gamma)$ known as resiter ("résidu itératif" in French) of the eigenradius γ and with residual series $R_i(y)$ that have in them only monomials of total degree $\geq$ p+1 in $y_0,y_1,\dots,y_{\nu-1}$ and of partial degree $\leq$ p-1 in y. Taken along with the level p, the resiter ρ and the directors λ_j, the coefficients*) appearing in the $R_i(y)$ constitute a complete system of formal invariants of the diffeo f.

*) These coefficients must be regarded as being defined upto a dilatation (2.6).

§3. HOLOMORPHIC INVARIANTS

The analytic classification turns out to be largely independent of the formal one. In fact, the only formal invariants that have a significant bearing on analytic classification are the level p, the resiter ρ, and the directors λ_j.

Let us (provisionally) maintain the genericity assumptions A_1 and A_2 and, to simplify further, let us start with the case $p = 1$, $\rho = 0$:

Proposition 2: (Formal integral. The case $p = 1$, $\rho = 0$).

For each eigenradius γ of resiter $\rho = 0$, the following system of difference equations:

$$x_i(z+1, u) = f_i(x_1(z,u),\dots,x_\nu(z,u)) \qquad (1 \le i \le \nu) \tag{3.1}$$

has a formal solution of type

$$x_i(z,u) \in \mathbb{C}[[z^{-1}, u_1\, z^{\lambda_1}, \dots, u_{\nu-1}\, z^{\lambda_{\nu-1}}]] \tag{3.2}$$

which is tangent to γ:

$$x_i(z,0) = \gamma_i\, z^{-1} + o(z^{-1}) \qquad \text{with} \qquad \gamma = (\gamma_1,\dots,\gamma_\nu) \quad . \tag{3.3}$$

This solution x(z,u) is known as the formal integral of f with respect to γ. It is unique up to translations on z and dilatations on u:

$$z \to z + c_0; \quad u_i \quad c_i\, u_i \to (c_i \in \mathbb{C}) \quad . \tag{3.4}$$

Proposition 3: (Resurgence of the formal integral. The case $p = 1$, $\rho = 0$).

Each of the formal integrals

$$x_i(z,u) = \sum_n u^n . \phi_i^n(z) \qquad (1 \le i \le \nu,\ u^n = u_1^{n_1} \dots u_{\nu-1}^{n_{\nu-1}}) \tag{3.5}$$

associated with any of the $\mathbb{N}$ eigenradii γ is resurgent in z (which simply means that its components $\phi_i^n(z)$ are resurgent in z). Moreover, it verifies resurgence equations that may be written compactly:

$$\Delta_\omega\, x(z,u) = A_\omega\, x(z,u) \qquad (\omega \in 2\pi i\, \mathbb{Z}) \tag{3.6}$$

with, on the left-hand side, alien derivations Δ_ω indexed on $2\pi i\, \mathbb{Z}$ and, on the right-hand side, ordinary differential operators whose coefficients are formal power series in u:

$$A_\omega = A^o_\omega(u)\frac{\partial}{\partial z} + \sum_{i=1}^{\nu-1} A^i_\omega(u)\frac{\partial}{\partial u_i} \tag{3.7}$$

$$A^o_\omega(u),\ A^1_\omega(u),\ldots,A^{\nu-1}_\omega(u) \in \mathbb{C}[[u_1,\ldots,u_{\nu-1}]] \ . \tag{3.8}$$

Proposition 4: (Holomorphic invariants)

The operators A_ω are holomorphic*) and analytic**) invariants of the diffeo f.

Proposition 5: (p and ρ arbitrary)

All the above results extend to the general case (p and ρ arbitrary) provided:

(i) equation (3.2) is replaced by:

$$x_i(z,u) \in \mathbb{C}[[z_o^{-1/p},\ u_1\, z_o^{\lambda_1/p},\ldots,u_{\nu-1}\, z_o^{\lambda_{\nu-1}/p}]] \tag{3.9}$$

with z_o deduced from z according to the law:

$$z = z_o + \rho \log z_o \tag{3.10}$$

(ii) the alien derivations Δ_ω appearing in (3.6) are no longer indexed on $2\pi i\,\mathbb{Z}^*$ but on the subset of $\mathbb{C}_p$ (meaning the Riemann surface of $z^{1/p}$) that lies over $2\pi i\,\mathbb{Z}^*$.

§4. ITERABILITY CRITERIA

Let us add to A_1 and A_2 a third assumption:

(A_3) Each eigenradius γ has (mutually) diophantian directors. In other words, the linear combinations $\sum n_i \lambda_i$ with integers n_i***) do not approximate zero faster than the "normal" speed. See [E2].

Observe that this simplifying hypothesis is "almost always" fulfilled. For instance, if the first Taylor coefficients of f are algebraic, so are the directors λ_i, and A_3 is automatically fulfilled.

---

*) Which means that they holomorphically depend on f, i.e. on its Taylor coefficients.

**) Which means that they are invariant under analytic changes of variables: $A_\omega(f) = A_\omega(h^{-1} \circ f \circ h)$

***) to which one must add an extra term $n_o\,\lambda_o = -n_o$.

Proposition 6: (Analytic conjugacy)

Under assumption A_3, the A_ω of propositions 3 and 4 constitute a full system of analytic invariants. In other words, two diffeos f and g tangent to identity are analytically conjugate if and only if they are formally conjugate and have the same holomorphic invariants A_ω.

It is common knowledge that diffeos tangent to identity possess formal iterates of any order (entire, rational, real, complex). Those iterates are unique (if one requires them to be also tangent to identity) but usually non-analytic.

Proposition 7: (Criteria for analytic iterability)

a) The set W(f) of all w such that f has an analytic iterate of order w is either $\mathbb{C}$ itself or a one-dimensional lattice of the form $q^{-1}\mathbb{Z}$ with q a positive integer.
b) $W(f) = \mathbb{C}$ iff all A_ω vanish
c) $W(f) = q^{-1}\mathbb{Z}$ iff all A_ω vanish except when ω lies in $2\pi iq\mathbb{Z}$. *)

With every eigenradius $\gamma = (\gamma_1,\ldots,\gamma_\nu)$ one may associate an entire orbit Γ of f, that is to say a formal curve:

$$\Gamma = \{\gamma_1(t),\ldots,\gamma_\nu(t)\} \qquad (\gamma_i \in \mathbb{C}[[t]]) \tag{4.1}$$

tangent to γ:

$$\gamma_i(t) = \gamma_i{}^t + o(t) \qquad (1 \leq i \leq \nu) \tag{4.2}$$

and formally invariant under f:

$$f_i(\gamma_1(t),\ldots,\gamma_\nu(t)) = \gamma_i(f^{\#}(t)) \qquad (f^{\#}(t) \in \mathbb{C}[[t]]) \quad . \tag{4.3}$$

Whenever it is possible to choose the parameter t in such a way that:

$$\gamma_1(t),\ldots,\gamma_\nu(t) \in \mathbb{C}\{t\} \tag{4.4}$$

Γ is said to be an analytic orbit.

*) Or over $2\pi iq\mathbb{Z}^*$ when the level p is ≥ 2.

Proposition 8: (Analytic orbits)

The entire orbit Γ associated with the eigenradius γ is analytic if and only if:

$$A_\omega^1(0) = A_\omega^2(0) = \ldots A_\omega^{\nu-1}(0) = 0 \qquad (\forall\omega) \tag{4.5}$$

for all the invariants A_ω associated with the eigenradius γ.

One observes that the first coefficient of A_ω, namely $A_\omega^0(u)$, does not intervene at all and that the remaining coefficients $A_\omega^1(u),\ldots,A_\omega^{\nu-1}(u)$ intervene only through their constant terms. Thus, the requirements for the existence of analytic orbits are much weaker than for analytic iterability.

§5. A PRIORI LINKS BETWEEN HOLOMORPHIC INVARIANTS

The holomorphic invariants A_ω are subject to two sorts of a priori constraints.

The first have to do with the growth properties in ω. They may be compactly written as follows:

$$(\exp(\sum_{m=1}^{+\infty} e^{\varepsilon 2\pi i m z} A_{\varepsilon 2\pi i m})).z \in \mathbb{C}\{e^{\varepsilon 2\pi i z}\} \text{ for } \varepsilon = \pm \quad . \tag{5.1}$$

The others have to do with the dependence in u of the A_ω. Indeed, it turns out that each A_ω is resurgent in each v_i *) and satisfies resurgence equations involving the corresponding A_ω associated with other eigenradii γ: The resurgence in v_i is of the "rigid" type. Its precise shape depends heavily on the sign of Re λ_i. See [E2].

§6. GENERAL DIFFEOMORPHISMS

The method we have just outlined extends, mutatis mutandis, to all local diffeos of $\mathbb{C}^\nu$, whether tangent to identity or not. It covers in particular the important case of local diffeos whose linear part has (multiplicatively) resonant eigenvalues. The resurgence equations (3.6) carry over to that and other cases, but with alien derivations indexed on sets richer than $2\pi i \mathbb{Z}$. The coefficients of the operators A_ω on the right-hand side of (3.6) always constitute complete systems of holomorphic invariants - and even of analytic invariants, except in rare instances when there occur parasitical phenomena such as nihilence or quasiresonance, which are responsible for the existence of the so-called "metaholomorphic" invariants (essentially non-constructible).

*) The v_i are parameters deduced from the u_i by putting $(v_i)^{\lambda_i} = (u_i)^{-1}$.

REFERENCES

[E1] Ecalle, J.: Les fonctions résurgentes, Volumes 1 et 2 (Pub. Math. d'Orsay, 1981)

[E2] Ecalle, J.: L'équation du pont et la classification analytique des objets locaux, to appear in Pub. Math. d'Orsay).

Jean Ecalle
Université de Paris Sud
Centre d'Orsay
Mathematique, Batiment 425
F-91405 Orsay Cedex

ON PSEUDO-PROCESSES AND THEIR EXTENSIONS

Maria Etgens and Zenon Moszner*

In [4] (see also [5] p.273) A.Pelczar gave the following definition of a pseudo-process. The definition is a generalization of a process introduced by C.M. Dafermos in [1,2].

Definition. Let Y be any non-empty set, (G;+) - an abelian semigroup with neutral element 0, H - a subsemigroup of G such that $0 \in H$ and let $\mu : G \times Y \times H \to Y$. The quadruplet (Y,G,H,μ) is called a pseudo-process if

$$\mu(t,y,0) = y \text{ and } \mu(t,y,u+v) = \mu(t+u,\mu(t,y,u),v)$$

for $y \in Y$, $t \in G$, $u,v \in H$.

In this paper we shall present the general form of a pseudo-process under the additional assumption on the subsemigroup H:

$$\bigwedge_{x \in G} \bigwedge_{u,v \in H} x + u = x + v \to u = v . \tag{1}$$

If (G;+) is an abelian group, we shall derive some necessary and sufficient conditions for the extension of a pseudo-process to a pseudo-process defined on $G \times Y \times G$.

The results given below are particular cases of more general considerations concerning the translation equation on Ehresmann groupoids (see [6]); however we shall carry out these considerations for facility of reading.

1. We assume the following definitions.

Definition 1. A family of sets $\{A_t\}_{t \in T}$ will be called a decomposition of the set A if

$$\bigwedge_{t \in T} A_t \neq \emptyset; \; A = \bigcup_{t \in T} A_t; \; \bigwedge_{t_1,t_2 \in T} t_1 \neq t_2 \to A_{t_1} \cap A_{t_2} = \emptyset .$$

Definition 2. (cf.[3]). The decomposition $\{B_s\}_{s \in S}$ of the set B will be called a subdecomposition of the decomposition $\{A_t\}_{t \in T}$ of the set A, if

$$\bigwedge_{s \in S} \bigwedge_{t \in T} B_s \subset A_t .$$

---

*Presented by Z.Moszner

Definition 3. (cf.[3]). The decomposition $\{B_s\}_{s \in S}$ of the set B will be called selective for the decomposition $\{A_t\}_{t \in T}$ of the set A, if

$$\bigwedge_{t \in T} \bigvee_{s \in S}^{1} B_s \subset A_t \quad .$$

Definition 4. A pseudo-process $(Y,G,G,\bar{\mu})$ is an extension on $G \times G$ of the pseudo-process (Y,G,H,μ), if $\mu = \bar{\mu}|_{G \times Y \times H}$.

Definition 5. A coset of the function $h : A \to B$ is said to be any set $h^{-1}(x)$ for $x \in h(A)$.

If A is a one-element set then we shall denote by $\}A\{$ the unique element of A.

2. For every $x \in G$ we shall denote by G_x the following set

$$G_x = \{t \in G : \bigvee_{u \in H} t + u = x\} .$$

Theorem 1. If

$$\mu(t,y,u) = \}h_{t+u}(h_t^{-1}(\{y\}))\{ \quad \text{for} \quad t \in G, \ \ y \in Y, \ \ u \in H , \tag{2}$$

where for $x \in G$ the functions h_x are the mappings of the sets $Y \times G_x$ onto the set Y, such that for every $t \in G$ and $u \in H$ the set of the cosets of the function h_t is a subdecomposition of the set of the cosets of the function h_{t+u}, then the quadruplet (Y,G,H,μ) is a pseudo-process.

Proof. The function μ is well-defined on the set $G \times Y \times H$. We have

$$\mu(t,y,0) = \}h_{t+0}(h_t^{-1}(\{y\}))\{ = \}h_t(h_t^{-1}(\{y\}))\{ = y$$

for $t \in G$, $y \in H$ and the neutral element 0.

Moreover, if $t \in G$, $y \in Y$ and $u,v \in H$, then

$$\mu(t+u,\mu(t,y,u,),v) = \}h_{(t+u)+v}(h_{t+u}^{-1}(h_{t+u}(h_t^{-1}(\{y\}))))\{ =$$

$$= \}h_{t+(u+v)}(h_t^{-1}(\{y\}))\{ = \mu(t,y,u+v) .$$

The mapping μ satisfies both conditions of the definition of a pseudo-process. This completes the proof.

The converse theorem to Theorem 1 is also true under the additional assumption (1).

Theorem 2. If condition (1) is fulfilled, then in every pseudo-process (Y,G,H,μ) the function μ has form (2), where h_x are functions satisfying the assumptions of Theorem 1.

Proof. We put

$$h_s(y,t) := \mu(t,y,u)$$

for $s,t \in G$, $y \in H$ and $u \in H$ such that $s = t + u$. Since condition (1) is fulfilled, the functions h_s are well-defined and the domain of each mapping h_s is the set

$$D_{h_s} = \{(y,t) \in Y \times G : \bigvee_{u \in H} t + u = s\} = Y \times G_s ,$$

for every $s \in G$.

Since for $y \in Y$ there exists $(y,s) \in Y \times G_s$ such that $h_s(y,s) = \mu(s,y,0) = y$, each function h_s transforms $Y \times G_s$ onto Y.

Let $y \in Y$, $t \in G$ and $u \in H$. We shall show that $h_t^{-1}(y) \subset h_{t+u}^{-1}(x)$ for $x = \mu(t,y,u)$, therefore, the set of the cosets of the function h_t is the subdecomposition of the set of the cosets of the function h_{t+u}. If $z,s \in h_t^{-1}(y)$, then $\mu(s,z,v) = y$ for $v \in H$ such that $s + v = t$. We then have

$$h_{t+u}(z,s) = h_{s+(v+u)}(z,s) = \mu(s,z,v+u) = \mu(s+v,\mu(s,z,v),u) =$$

$$= \mu(t,y,u) = x .$$

Now we shall prove that equality (2) holds. For $(t,y,u) \in G \times Y \times H$ we have $\}h_{t+u}(h_t^{-1}(\{y\}))\{ = h_{t+u}(y,t)$ since $h_t(y,t) = \mu(t,y,0) = y$. But $h_{t+u}(y,t) = \mu(t,y,u)$.

3. We shall consider the problem of the extendability of the pseudo-process in the case where $(G;+)$ is an abelian group.

Theorem 3. The pseudo-process (Y,G,H,μ), where $(G;+)$ is an abelian group, can be extended on the set $G \times Y \times G$ if and only if there exists a decomposition R of the set $Y \times G$ such that, for every $t \in G$, the set of the cosets of the function h_t occuring in condition (2) is a selective decomposition for the decomposition R.

Proof. Let us consider the decomposition $\{R_s\}_{s \in S}$ of the set $Y \times G$ satisfying the assumptions of Theorem 3. Since the function μ can be defined by condition (2), where h_t are the functions mapping $Y \times G_t$ onto Y, we can put

$$\bar{h}_t(y,v) := x \Leftrightarrow (y,v) \in R_s \supset h_t^{-1}(x)$$

for $t \in G$ and $(y,v) \in Y \times G$.

Then for every $t \in G$ the function $\bar{h}_t$ transforms $Y \times G$ onto Y, and, moreover, the set of the cosets of the function $\bar{h}_t$ is just the same as the set $\{R_s\}_{s \in S}$. The conditions of the Theorem 1 are satisfied, therefore the quadruplet $(Y,G,G,\bar{\mu})$, where $\bar{\mu}$ is defined by (2) with the parameters $\bar{h}_t$, is a pseudo-process.

We have $\bar{h}_t(y,v) = h_t(y,v)$ for $(y,v) \in Y \times G_t$. Hence $\bar{h}_t(y,t) = h_t(y,t)$ therefore $\bar{h}_t(y,t) = y$ so $(y,t) \in \bar{h}_t^{-1}(y)$. Moreover, for $(t,y,u) \in G \times Y \times H$ we get

$$\bar{\mu}(t,y,u) = \rbrace\bar{h}_{t+u}(\bar{h}_t^{-1}(\{y\}))\lbrace = \bar{h}_{t+u}(y,t) = h_{t+u}(y,t) =$$

$$= \rbrace h_{t+u}(h_t^{-1}(\{y\}))\lbrace = \mu(t,y,u) ,$$

because $(y,t) \in G_{t+u}$.

Now let us assume that $(Y,G,G,\bar{\mu})$ is an extension of (Y,G,H,μ) onto $G \times G$. Let ρ be a relation on $Y \times G$ defined as follows

$$(y,t)\rho(x,s) \Leftrightarrow \bar{\mu}(t,y,-t+s) = x \quad .$$

Since $\bar{\mu}(t,y,0) = y$, ρ is reflexive. If $\bar{\mu}(t,y,-t+s) = x$ then

$$\bar{\mu}(s,x,-s+t) = \bar{\mu}(t+(-t+s),\bar{\mu}(t,y,-t+s),-s+t) = \bar{\mu}(t,y,0) = y ,$$

so ρ is symmetric. Let $(y,t)\rho(x,s)$ and $(x,s)\rho(z,r)$. Then

$$z = \bar{\mu}(s,\bar{\mu}(t,y,-t+s),-s+r) = \bar{\mu}(t,y,-t+r) ,$$

and so $(y,t)\rho(z,r)$, therefore ρ is transitive.

We put $R := Y \times G/\rho$. It is necessary to prove that for every $t \in G$ the set of the cosets of the function h_t is a selective decomposition of the set $Y \times G_t$ for the decomposition $Y \times G$. For this purpose it is sufficient to prove that the following conditions:

$1^0 \qquad Y \times G_{t/\rho} = \{h_t^{-1}(\{y\})\}_{y \in Y}$,

$2^0 \qquad \bigwedge_{(x,s) \in Y \times G} h_t^{-1}(\bar{\mu}(s,x,-s+t)) \cap [(x,s)]_\rho \neq \emptyset$

hold for $t \in G$.

Let (x,s), (z,t) be arbitrary pairs in the set $Y \times G_t$. Then there exist $u,v \in H$ such that $s + u = t$ and $r + v = t$ and

$$(x,s)\rho(z,r) \Leftrightarrow \bar{\mu}(s,x,-s+r) = z \Leftrightarrow \bar{\mu}(r,\bar{\mu}(s,x,-s+r),-r+s+u) =$$

$$= \bar{\mu}(r,z,-r+s+u) \Leftrightarrow \bar{\mu}(s,x,u) = \bar{\mu}(r,z,-r+t) \Leftrightarrow$$

$$\Leftrightarrow \bar{\mu}(s,x,u) = \bar{\mu}(r,z,v) \Leftrightarrow \mu(s,x,u) = \mu(r,z,v) \Leftrightarrow$$

$$\Leftrightarrow \}h_t(h_s^{-1}(\{x\}))\{ = \}h_t(h_r^{-1}(\{z\}))\{ \Leftrightarrow h_t(x,s) = h_t(z,r) \; .$$

We have proved that equality 1^0 holds.

Now it is enough to notice that $(y,t) \in h_t^{-1}(\{y\})$ for an arbitrary $y \in Y$ and moreover

$$\bar{\mu}(t,\bar{\mu}(s,x,-s+t),-t+s) = \bar{\mu}(s,x,0) = x \; .$$

Therefore $(\bar{\mu}(s,x,-s+t),t)\rho(x,s)$, which completes the proof.

An additional assumption enables us to obtain the following

Theorem 4. If $(G;+)$ is an abelian group and H is a subsemigroup of G such that $0 \in H$ and $H \cup (-H) = G$, where $-H := \{t \in G : -t \in H\}$, then the pseudo-process (Y,G,H,μ) can be extended onto the set $G \times G$ if and only if the function $\mu(t,.,u)$ is a bijection from Y onto itself for every $t \in G$ and $u \in H$.

If the above condition is fulfilled, then the pseudo-process $(Y,G,G,\bar{\mu})$, where

$$\bar{\mu}(t,y,u) = \begin{cases} \mu(t,y,u) & \text{for} \quad t \in G, \quad y \in Y, \quad u \in H, \\ \mu^{-1}(t+u,.,-u)(y) & \text{for} \quad t \in G, \quad y \in Y, \quad u \in (-H), \end{cases} \tag{3}$$

is the unique extension of (Y,G,H,μ).

Proof. Let $(Y,G,G,\bar{\mu})$ be an arbitrary extension of the pseudo-process (Y,G,H,μ) on the set $G \times G$. Let us assume that $\mu(t,x,u) = \mu(t,y,u)$ for $(t,u) \in G \times H$ and $x,y \in Y$. Then

$$x=\bar{\mu}(t,x,0)=\bar{\mu}(t+u,\bar{\mu}(t,x,u),-u)=\bar{\mu}(t+u,\bar{\mu}(t,y,u),-u)=\bar{\mu}(t,y,0) = y \; .$$

Moreover $\mu(t,\bar{\mu}(t+u,y,-u),u)=\bar{\mu}(t+u,y,0) = y$ for any $y \in Y$. Thus $\mu(t,.,u)$ is bijection of Y onto itself.

Let us assume that the functions $\mu(t,.,u)$ are bijections of the set Y onto Y. We shall show that the function $\bar{\mu}$ defined by (3) is the extension of μ. The function $\bar{\mu}$ is well-defined on the set $G \times Y \times G$. Indeed, if $u,-u \in H$, then

$$\mu(t+u,\mu(t,y,u),-u) = \mu(t,y,0) = y\ ,$$

thus $\mu^{-1}(t+u,.,-u)(y) = \mu(t,y,u)$. It follows from the definition of the function $\bar{\mu}$ that $\bar{\mu}|_{G \times Y \times H} = \mu$.

We shall verify that $\bar{\mu}$ satisfies the second condition of the definition of the pseudo-process. The following cases are possible:

a) $u \in H$, $v \in (-H)$, $u + v \in H$,

b) $u \in H$, $v \in (-H)$, $u + v \in (-H)$,

c) $u \in (-H)$, $v \in (-H)$, $u + v \in (-H)$,

d) $u \in (-H)$, $v \in H$, $u + v \in H$,

e) $u \in (-H)$, $v \in H$, $u + v \in (-H)$,

f) $u \in H$, $v \in H$, $u + v \in H$.

In case a) we have

$$\bar{\mu}(t+u,\bar{\mu}(t,y,u),v) = [\mu^{-1}(t+u+v,.,-v)](\mu(t,y,u)) =$$

$$= [\}h_{t+u}(h^{-1}_{t+u+v}(.))\{]^{-1}(\}h_{t+u}(h^{-1}_{t}(\{y\}))\{) =$$

$$= \}h_{t+u+v}(h^{-1}_{t+u}(.))\{(\}h_{t+u}(h^{-1}_{t}(\{y\}))\{) = \}h_{t+u+v}(h^{-1}_{t}(\{y\}))\{ =$$

$$= \mu(t,y,u+v) = \bar{\mu}(t,y,u+v).$$

In case b) we have

$$\bar{\mu}(t+u,\bar{\mu}(t,y,u),v) = [\mu^{-1}(t+u+v,.,-v)](\mu(t,y,u)) = \}h_{t+u+v}(h^{-1}_{t}(\{y\}))\{ =$$

$$= [\}h_{t}(h^{-1}_{t+u+v}(.))\{]^{-1}(y) = \mu^{-1}(t+u+v,.,-(u+v))(y) = \bar{\mu}(t,y,u+v).$$

In case c) we have

$$\bar{\mu}(t+u,\bar{\mu}(t,y,u),v) = [\mu^{-1}(t+u+v,.,-v)](\mu^{-1}(t+u,.,-u)(y)) =$$

$$= [\}h_{t+u}(h^{-1}_{t+u+v}(.))\{]^{-1}([\}h_t(h^{-1}_{t+u}(.))\{]^{-1}(y)) =$$

$$= [\}h_{t+u+v}(h^{-1}_{t+u}(.))\{]^{-1}([\}h_{t+u}(h^{-1}_{t}(.))\{](y) = \}h_{t+u+v}(h^{-1}_{t}(.))\{(y) =$$

$$= [\}h_t(h^{-1}_{t+u+v}(.)\{]^{-1}(y) = \mu^{-1}(t+u+v,.,-(u+v))(y) = \bar{\mu}(t,y,u+v) .$$

The proofs of the cases d) and e) are analogous and case f) is evident. Thus $(Y,G,G,\bar{\mu})$ is an extension of the pseudo-process (Y,G,H,μ).

Now we shall prove that the pseudo-process $(Y,G,G,\bar{\mu})$, where $\bar{\mu}$ is defined by (3), is the unique extension of (Y,G,H,μ). Let (Y,G,G,μ_1), (Y,G,G,μ_2) be arbitrary extensions of the pseudo-process (Y,G,H,μ). For $t \in G$, $y \in Y$ and $u \in H$ we have

$$\mu(t+u,\mu_1(t,y,-u),u) = \mu(t,y,0) = \mu(t+u,\mu_2(t,y,-u),u) .$$

Since the mapping $\mu(t+u,.,u)$ is a bijection of Y onto itself, $\mu_1(t,y,-u)=\mu_2(t,y,-u)$. From $H \cup (-H) = G$ it follows that $\mu_1 = \mu_2$, which completes the proof.

Remark. The extendability of the pseudo-process (Y,G,H,μ) can be proved on the ground of Theorem 3. It is sufficient to define a relation ρ on the family of all cosets of the function h_x for $x \in G$ as follows

$$W_1 \,\rho\, W_2 \Leftrightarrow W_1 \subset W_2 \quad \text{or} \quad W_2 \subset W_1 .$$

If the functions $\mu(t,.,u)$ are bijections of Y onto Y, then ρ is an equivalence relation. The components of the decomposition R of the set $Y \times G$ are the unions of all equivalent cosets. The decomposition R satisfies the conditions of Theorem 3.

We shall illustrate the above results in the following examples.

Example 1. Let Y be the set R^+ of all positive real numbers, $(G;+)$ - the group $(R;+)$ of real numbers under addition, and H - the subsemigroup R^+_0 of all non-negative real numbers. Then for $t \in R$

$$G_t = \{x \in R : \bigvee_{u \in R^+_0} x = t - u\} .$$

We put for $y \in R^+$ and $t - u \in G_t$

$$h_t(y,t-u) = y\, e^u .$$

For an arbitrary $x \in R^+$ we have $h_t(x\, e^{-u}, t-u) = x\, e^{-u} e^u = x$, thus the function h_t transforms $R^+ \times G_t$ onto R^+. Since for $y \in R^+$

$$h_t^{-1}(\{y\}) = \{(x,t-u) \in R^+ \times G_t : x = y\, e^{-u} \quad \text{and} \quad u \in R_o^+\}$$

moreover

$$h_{t+v}(y\, e^{-u},t-u) = h_{t+v}(y\, e^{-u},t+v-(u+v)) = y\, e^{-u}\, e^{u+v} = y\, e^v ,$$

we have $h_t^{-1}(y) \subset h_{t+v}^{-1}(y\, e^v)$ for any $t \in R$, $v \in R_o^+$. Therefore, the set $\{h_t^{-1}(\{y\})\}_{y \in R^+}$ is a subdecomposition of the decomposition $\{h_{t+v}^{-1}(\{y\})\}_{y \in R^+}$. In virtue of Theorem 1 it follows that the quadruplet (R,R^+,R_o^+,μ), where $\mu(t,y,v) = \,]h_{t+v}(h_t^{-1}(\{y\}))[\, = h_{t+v}(y\, e^{-u},t-u) = y\, e^v$, for $(t,y,v) \in R \times R^+ \times R_o^+$, is a pseudo-process. This pseudo-process can be extended on $R \times R$, because for each $(t,v) \in R \times R_o^+$ the function $\mu(t,.,v)$ is a bijection of R^+ onto R^+. It follows from Theorem 4 that the pseudo-process $(R,R^+,R,\bar{\mu})$, where $\bar{\mu}(t,y,u) = y\, e^u$, is its unique extension.

<u>Example 2.</u> Let $(G;+)$ be an abelian group such that card $G > 1$ and let H be a sub-semigroup such that $H \cup (-H) = G$ and $H \cap (-H) = \{0\}$. Moreover, let us assume that $Y = G$ and put

$$\mu(t,y,u) = \begin{cases} t + u & \text{for} \quad t,y \in G \quad \text{and} \quad u \in H - \{0\}, \\ y & \text{for} \quad t,y \in G \quad \text{and} \quad u = 0 . \end{cases}$$

For $t,y \in G$ and $u,v \in H - \{0\}$ we have

$$\mu(t+u,\mu(t,y,u),v) = (t+u) + v = t + (u+v) = \mu(t,y,u+v) .$$

If $t,y \in G$, $u = 0$ and $v \in H - \{0\}$ then

$$\mu(t+0,\mu(t,y,0),v) = t + v = \mu(t,y,0+v) .$$

However, for $t,y \in G$, $u \in H - \{0\}$ and $v = 0$ we get

$$\mu(t+u,\mu(t,y,u),0) = \mu(t,y,u) = t + u = \mu(t,y,u+0) .$$

In the case, when $t,y \in G$ and $u = v = 0$ we obtain

$$\mu(t+0,\mu(t,y,0),0) = \mu(t,y,0) = y = \mu(t,y,0+0) .$$

Thus the quadruplet $(G,G,H,\mu)$is a pseudo-process. This pseudo-process cannot be extended to $G \times G$ because for all $t,y \in G$ and $u \in H - \{0\}$ the equality

$$y = \mu(t,y,0) = \mu(t-u,\bar{\mu}(t,y,-u),u) = (t - u) + u = t$$

does not hold.

This is conform with Theorem 4 because the function $\mu(t,.,u)$ for $(t,u) \in G \times$ $\times (H - \{0\})$ is not a bijection on the set G.

4. A pseudo-dynamical semi-system (cf. 5 , p.17) defined as the triplet (Y,H,π), where Y is an arbitrary non-empty set, H - an abelian semigroup with neutral element 0, π - a function transforming $H \times Y$ into Y and satisfying the following conditions

$$\pi(0,y) = y \quad \text{for every} \quad y \in Y ,$$

$$\pi(s,\pi(t,y)) = \pi(t+s,y) \quad \text{for all} \quad y \in Y \quad \text{and} \quad s,t \in H , \tag{4}$$

is a particular case of the pseudo-process in which the function is independent of the first variable. We can use the general form (2) of the function μ in a pseudo-process to get a general form of the function π in a pseudo-dynamical semi-system. For this purpose, it is necessary and sufficient to define the family of functions h_x for $x \in G$ fulfilling the assumptions of Theorem 1 such that the function

$$\}h_{t+u}(h_t^{-1}(\{y\}))\{$$

is independent of t. This means that it is sufficient to determine the solutions of equations (4). Then the function $\pi(y,u) = \}h_u(h_0^{-1}(\{y\}))\{$ must satisfy the equality (4). Accepting some assumptions on G and H (e.g. $G = H = R^+$) equation (4) must be satisfied by the function $h_u(y)$, immediately.

REFERENCES

[1] Dafermos, C.M.: An invariance principle for compact processes, J.Diff.Eqs.9, 239-252 (1971).

[2] Dafermos, C.M.: Applications of the invariance principle for compact processes, I. Asymptotically dynamical systems, ibid. 291-299.

[3] Grząślewicz, A.: On extensions of homomorphisms, Aequationes Math. 17 No.2/3, 199-207 (1978).

[4] Pelczar, A.: Stability questions in generalized processes and in pseudo-dynamical systems, Bull. Acad. Polon. Sci. Sér. Sci. Math., Astronom., Phys. 21, 541-549 (1973).

[5] Pelczar, A.: A general dynamical systems, script U.J., Kraków, 1978.

[6] Żurek-Etgens, M.: Systemy odwrotne a równanie translacji (The inverse systems and the translation equation), typescript of doctor's thesis, 1982.

Maria Etgens, Zenon Moszner
Kazimierza Wielkiego 87/8
PL 30-074 Kraków, Poland

THE PILGERSCHRITT TRANSFORM IN LIE ALGEBRAS

Wolfgang Förg-Rob

INTRODUCTION

Let G be a Lie group and L = L(G) its Lie algebra. If $\phi:[0,1] \to G$ is a C^1-path with $\phi(0) = e$ (unit element of G), then its pilgerschritt transform $\tilde{\phi}$ can be defined by the following rule: Solve the equations

$$Y_\tau'(t) = \tau\phi'(t)\phi(t)^{-1}Y_\tau(t), \quad Y_\tau(0) = e$$

and put $\hat{\phi}(t,\tau) := Y_\tau(t)$ for each $\tau \in [0,1]$.
Then $\tilde{\phi}$ is given by $\tilde{\phi}(t) := \hat{\phi}(1,t)$.
It has been shown that the sequence $\phi, \tilde{\phi}, \tilde{\tilde{\phi}}, \tilde{\tilde{\tilde{\phi}}}, \ldots$ converges to the restriction of a one-parameter subgroup $h:\mathbb{R} \to G$ with the property $h(1) = \phi(1)$ under certain conditions on ϕ or G ([2],[4]), but this sequence may also diverge or oscillate ([3]).

In this paper it will be proved that such a behaviour of the sequence $\phi, \tilde{\phi}, \tilde{\tilde{\phi}}, \ldots$ does not depend on G and ϕ, but only on L and $f:[0,1] \to L:t \to \phi'(t)\phi(t)^{-1}$.

Furthermore, a simple proof for the convergence of the sequence $\phi, \tilde{\phi}, \tilde{\tilde{\phi}}, \ldots$ under certain conditions will be given.

A second part of this paper deals with the pilgerschritt transform in the group of vectors of formal power series. The methods used in the first part will be used to prove some results on the behaviour of the pilgerschritt transform in this group.

Results. Now let $\phi:[0,1] \to G$, $\phi(0) = e$, be a given C^1-path. Usual existence and uniqueness-theorems on differential equations show that the correspondence

$$\{\psi:[0,1] \to G \mid \psi(0) = e, \psi \in C^1\} \to \{g:[0,1] \to L \mid g \in C^0\}$$
$$\psi \to (t \to \psi'(t)\psi(t)^{-1})$$

is one-to-one and onto. Thus the proposition stated in the introduction may be formulated as the following theorem:

Theorem 1. Let $f:[0,1] \to L:t \to \phi'(t)\phi(t)^{-1}$. Then $\tilde{f}(t) := \frac{\partial\tilde{\phi}}{\partial t}(t)\tilde{\phi}(t)^{-1}$ can be computed within L only (without any calculations in the group G) from f.

Proof. We have

$$\tilde{f}(\tau) = \frac{\partial\tilde{\phi}}{\partial\tau}(\tau)\tilde{\phi}(\tau)^{-1} = \frac{\partial\hat{\phi}}{\partial\tau}(1,\tau)\hat{\phi}(1,\tau)^{-1} = \mathrm{Ad}(\hat{\phi}(1,\tau))(\hat{\phi}(1,\tau)^{-1}\cdot\frac{\partial\hat{\phi}}{\partial\tau}(1,\tau))$$

1) $\Phi(t,\tau)^{-1}\cdot \frac{\partial\Phi}{\partial\tau}(t,\tau) \in L$ for each $t,\tau \in [0,1]$. As L is a Banach space, we may identify L and its tangent space, and therefore $\frac{\partial}{\partial t}(\Phi(t,\tau)^{-1}\cdot\frac{\partial\Phi}{\partial\tau}(t,\tau))$ is an element of L. By the usual identifications in the theory of Lie groups we get

$$\frac{\partial}{\partial t}(\Phi(t,\tau)^{-1}\cdot \frac{\partial\Phi}{\partial\tau}(t,\tau)) =$$

$$= -\Phi(t,\tau)^{-1}\cdot \frac{\partial\Phi}{\partial t}(t,\tau)\cdot\Phi(t,\tau)^{-1}\cdot \frac{\partial\Phi}{\partial\tau}(t,\tau) + \Phi(t,\tau)^{-1}\cdot \frac{\partial^2\Phi}{\partial t\partial\tau}(t,\tau) =$$

$$= \Phi(t,\tau)^{-1}\cdot f(t)\cdot\Phi(t,\tau) =$$

$$= \mathrm{Ad}(\Phi(t,\tau))^{-1}(f(t)) \quad .$$

As $\Phi(0,\tau) = e$ for each τ, we get $\frac{\partial\Phi}{\partial\tau}(0,\tau) = 0$ and therefore

$$\Phi(t,\tau)^{-1}\cdot \frac{\partial\Phi}{\partial\tau}(t,\tau) = \int_0^t \mathrm{Ad}(\Phi(\xi,\tau))^{-1}(f(\xi))d\xi \ .$$

Furthermore,

$$\mathfrak{f}(\tau) = \mathrm{Ad}(\Phi(1,\tau))(\int_0^1 \mathrm{Ad}(\Phi(\xi,\tau))^{-1}(f(\xi))d\xi) \quad . \tag{*}$$

2) For each $t,\tau \in [0,1]$ the map $\mathrm{Ad}(\Phi(t,\tau))$ is a linear map from L to L. As $\mathrm{Ad}:G \to \mathrm{Gl}(L)$ is a Lie group homomorphism, $\mathrm{Ad}(\Phi(t,\tau))$ satisfies the differential equation

$$\frac{\partial}{\partial t}\mathrm{Ad}(\Phi(t,\tau)) = \mathrm{ad}(\frac{\partial\Phi}{\partial t}(t,\tau)\cdot\Phi(t,\tau)^{-1})\cdot\mathrm{Ad}(\Phi(t,\tau))$$

$$= \mathrm{ad}(\tau f(t))\cdot\mathrm{Ad}(\Phi(t,\tau))$$

$$\mathrm{Ad}(\Phi(0,\tau)) = \mathrm{Ad}(e) = \mathrm{Id}_L \quad .$$

Thus $\mathrm{Ad}(\Phi(t,\tau))$ is the unique solution of the differential equation

$$A_\tau'(t) = \mathrm{ad}(\tau f(t))\cdot A_\tau(t) \ , \qquad A_\tau(0) = \mathrm{Id}_L$$

in $\mathrm{Gl}(L)$, and this solution can be computed within $\mathrm{Gl}(L)$, which completes the proof.

Corollary. The proof of theorem 1 shows that $\mathfrak{f}(\tau)$ is an analytic function in the argument τ.

In the sequel we use the notation $A_\tau(t)$ instead of $\mathrm{Ad}(\Phi(t,\tau))$, because $A_\tau(t)$ does not depend on ϕ, but only on f.

Proposition 1. Let $L = L^0 \supset L^1 \supset L^2 \supset \ldots$ be the lower central series of L, and $f:[0,1] \to L$ be C^1. If $f'(t) \in L^n$ for all t, then $\tilde{f}'(\tau) \in L^{n+1}$, and the equation

$$\tilde{f}'(\tau) = A_\tau(1)\cdot\int_0^1\int_\xi^1\int_\xi^\mu [A_\tau(\eta)^{-1}f'(\eta),\ A_\tau(\xi)^{-1}f(\xi)]\,d\eta\, d\mu\, d\xi$$

holds for all τ.

Proof. As $A_\tau(t)$ is the solution of the equation

$$A_\tau'(t) = \tau\mathrm{ad}(f(t))\cdot A_\tau(t),\quad A_\tau(0) = \mathrm{Id}_L$$

we get

$$\frac{\partial}{\partial\tau} A_\tau(t) = \int_0^t A_\tau(t)\cdot A_\tau(\xi)^{-1}\cdot \mathrm{ad}(f(\xi))\cdot A_\tau(\xi)d\xi =$$

$$= A\ (t)\cdot\int_0^t \mathrm{ad}(A_\tau(\xi)^{-1}\cdot f(\xi))d\xi\ .$$

Furthermore,

$$\frac{\partial}{\partial\xi}(A_\tau(\xi)^{-1}\cdot f(\xi)) = A_\tau(\xi)^{-1}\cdot f'(\xi) \quad \text{and}$$

$$A_\tau(\mu)^{-1}\cdot f(\mu) = A_\tau(\xi)^{-1}\cdot f(\xi) + \int_\xi^\mu A_\tau(\eta)^{-1}\cdot f'(\eta)d\eta\ .$$

Differentiation of equation (*) leads to the desired equation for $\tilde{f}'$; as $A_\tau(t)$ leaves L^k invariant, the proof is complete.

Corollary 1. Consider the sequence $f_1 = f$, $f_{n+1} = \tilde{f}_n$. If L is nilpotent of degree N, then $f_N' = 0$, thus f_N is a constant function.

Corollary 2. If f is constant, then $f' = 0$, and therefore $\tilde{f} = f$.

Now we can state a theorem on convergence of the sequence of iterated pilgerschritt transforms:

Theorem 2. Assume that $\|[-,*]\|_{op} = M$, and $\|f(0)\| + \sup_{t\in[0,1]}\|f'(t)\| \le x_o$, where x_o is the positive solution of the equation

$$2(Mx)^2 = e^{2Mx}(2Mx-3) + 4e^{Mx}-1 \text{ (at about } Mx_o = 0.91) \quad .$$

Then we also have $\|\tilde{f}(0)\| + \sup_{t\in[0,1]}\|\tilde{f}'(t)\| \le x_o$, and

$$\sup_{t\in[0,1]}\|\tilde{f}'(t)\| \le \frac{1}{2}\cdot \sup_{t\in[0,1]}\|f'(t)\|\ .$$

The sequence $f_1 = f$, $f_{n+1} = \tilde{f}_n$ converges uniformly to a constant function.

Proof. Let $t \in [0,1]$, then

$$\|f(t)\| = \|f(0) + \int_0^t f'(\xi)d\xi\| \leq$$

$$\leq \|f(0)\| + \sup_{\xi \in [0,1]} \|f'(\xi)\| \leq x_0$$

thus $\|ad(\tau f(t))\|_{op} \leq Mx_0$ for each $t, \tau \in [0,1]$ and therefore

$$\|A_\tau(1) \cdot A_\tau(\xi)^{-1}\|_{op} \leq e^{(1-\xi)Mx_0} \quad \text{for each } \xi \in [0,1] .$$

This leads to

$$\|\tilde{f}'(\tau)\| \leq \int_0^1 \int_\xi^1 \int_\xi^\mu M \cdot e^{(1-\eta)Mx_0} \cdot e^{(1-\xi)Mx_0} d\eta d\mu d\xi \; x_0 \cdot \sup_{t \in [0,1]} \|f'(t)\|$$

and further

$$\sup_{\tau \in [0,1]} \|\tilde{f}'(\tau)\| \leq \sup_{t \in [0,1]} \|f'(t)\| \cdot \frac{e^{2Mx_0}(2Mx_0 - 3) + 4e^{Mx_0} - 1}{4M^2 x_0^2} =$$

$$= \sup_{t \in [0,1]} \|f'(t)\| \cdot \frac{1}{2} .$$

As $A_0(t) = Id_L$, we get

$$\tilde{f}(0) = \int_0^1 f(t)dt = f(0) + \int_0^1 (1-\xi) f'(\xi)d\xi , \quad \text{thus}$$

$$\|\tilde{f}(0)\| + \sup_{\tau \in [0,1]} \|\tilde{f}'(\tau)\| \leq$$

$$\leq \|f(0)\| + \frac{1}{2} \cdot \sup_{\xi \in [0,1]} \|f'(\xi)\| + \frac{1}{2} \cdot \sup_{t \in [0,1]} \|f'(t)\| \leq x_0 .$$

The estimate above shows that f_n' converges to 0 uniformly, which implies the uniform convergence of f_n to a constant function.

Corollary. Let $\phi: [0,1] \to G$, $\phi(0) = e$, be a C^2-path such that $f(t) := \phi'(t)\phi(t)^{-1}$ fulfills the conditions of theorem 2. If we write $\phi_1 = \phi$, $\phi_{n+1} = \tilde{\phi}_n$, then ϕ_n is the solution of the differential equation

$$\psi'(t) = f_n(t)\psi(t), \quad \psi(0) = e,$$

and a proposition in [1] guarantees that the sequence ϕ_n converges uniformly to the restriction of a one-parameter-subgroup.

Part 2. Let $L = (\mathbb{R}[[x]])^n$ or $L = (\mathbb{C}[[x]])^n$ be the vector space of n-vectors of formal power series (FPS) in the indeterminates $x = (x_1,\dots,x_n)$. We write $F(x) = \sum_{i\geq 1} F_i(x,\dots,x) = \sum_{i\geq 1} F_i(x)$, where each F_i is a homogeneous polynomial of degree i with values in $\mathbb{R}^n$ resp. $\mathbb{C}^n$. The composite FPS $F \circ G$ of two FPS F, G is given by

$$F \circ G(x) = \sum_{k\geq 1} \sum_{1\leq p\leq k} \sum_{i_1+\dots+i_p=k} F_p \circ (G_{i_1},\dots,G_{i_p})(x) .$$

The set $G = \{F \in L | F_1 \text{ is invertible}\}$ forms a group under this composition, and the inverse H of a FPS $F \in G$ is given by

$$H_1 = (F_1)^{-1}$$

$$H_k = -F_1^{-1}(\sum_{2\leq p\leq k} \sum_{i_1+\dots+i_p=k} F_p \circ (H_{i_1},\dots,H_{i_p}))$$

recursively.

Furthermore, we will use the following notations: For $F \in L$, $F(x) = \sum_{i\geq 1} F_i(x)$, let $TF(x,y)$ and $T^2F(x;y,z)$ denote the FPS

$$TF(x,y) = F_1(y) + \sum_{k\geq 2} \sum_{1\leq i\leq k} F_i(x,\dots,\underset{\substack{\uparrow \\ k\text{-th component}}}{y},\dots x)$$

$$T^2F(x;y,z) = F_2(y,z) + F_2(z,y) + \sum_{k\geq 3} \sum_{\substack{1\leq k_1,k_2\leq k \\ k_1\neq k_2}} F_k(x,\dots,\underset{\substack{\uparrow \\ k_1\text{-th component}}}{y},\dots,\underset{\substack{\uparrow \\ k_2\text{-th component}}}{z},\dots x) .$$

It is easy to see that TF, T^2F can be treated like derivatives of F, e.g., the chain rule holds: Let F,G be FPS, then $T(F \circ G)(x,y) = TF(G(x),TG(x,y))$.

If $F:[0,1] \to L:t \to \sum_{i\geq 1} F_i(t)(x)$ is a family of FPS, and if all the coefficients F_i are differentiable at $t \in [0,1]$, then we denote by $F'(t)(x)$ the formal power series

$$F'(t)(x) = \sum_{i\geq 1} F_i'(t)(x) .$$

After these remarks on the notation we give a first lemma:

Proposition 2. Let $f:[0,1] \to L:t \to \sum_{i\geq 1} F_i(t)(x)$ be given such that the coefficients $F_i(t)$ are continuous in the argument t. Then the differential equation $G'(t)(x) = F(t) \circ G(t)(x)$ has a uniquely determined solution for the initial value $G(0)(x) = x$. Furthermore, $G(t)(x)$ is invertible for each t, and we have $F(t) = G'(t) \circ G(t)^{-1}$.

Proof. For the homogeneous components, the differential equation is given by

$$G_1'(t) = F_1(t)\cdot G_1(t) \qquad G_1(0) = \mathrm{Id}$$

$$G_k'(t) = F_1(t)\cdot G_k(t) + \sum_{2\leq p\leq k} \;\sum_{i_1+\ldots+i_p=k} F_p(t)\cdot(G_{i_1}(t),\ldots,G_{i_p}(t))$$

$$G_k(0) = 0 \quad \text{for} \quad k \geq 2 \quad .$$

Thus the proposition is a trivial consequence of usual theorems on ordinary linear differential equations.

One of the main results of part 2 is the fact that the pilgerschritt transform can be used to compute the coefficients of a one-parameter-subgroup through a given point $\phi(1)(x)$ in G under the condition that $\phi(t)_1(x) = x$ for all t.

We define the pilgerschritt transform in the group G in the following way - analogous to the definition of the pilgerschritt transform in Lie groups:

Definition 1. Let $\phi:[0,1] \to G$ be a path such that the coefficients are C^1 and $\phi(0)(x) = x$. Define $F:[0,1] \to L$ by $F(t) := \phi'(t) \circ \phi(t)^{-1}$ and solve the equation

$$G_\tau'(t) = \tau.F(t) \circ G_\tau(t), \quad G_\tau(0)(x) = x$$

for each $\tau \in [0,1]$. Put $\hat{\phi}(t,\tau) := G_\tau(t)$ and call $\tilde{\phi}(\tau) := \hat{\phi}(1,\tau)$ the pilgerschritt transform of ϕ.

Proposition 2 shows that $\tilde{\phi}$ exists and is uniquely determined. We denote by

$$\tilde{F}(\tau) := \frac{\partial\tilde{\phi}}{\partial\tau}(\tau) \circ \tilde{\phi}(\tau)^{-1}$$

the pilgerschritt transform of F, and

$$\hat{F}(t,\tau) := \frac{\partial\hat{\phi}}{\partial\tau}(t,\tau) \circ \hat{\phi}(t,\tau)^{-1}$$

$$H(t,\tau) := T\hat{\phi}(t,\tau)^{-1}\left(\hat{\phi}(t,\tau), \frac{\partial\hat{\phi}}{\partial\tau}(t,\tau)\right) \quad .$$

Proposition 3. Let ϕ, F, Φ be as in definition 1. Then $\frac{\partial\Phi}{\partial\tau}$ is given by

$$(^{**})\quad \frac{\partial\Phi}{\partial\tau}(t,\tau)(x)=T\Phi(t,\tau)(x,\int_0^t T\Phi(\xi,\tau)^{-1}(\Phi(\xi,\tau)(x),F(\xi)\circ\Phi(\xi,\tau)(x))d\xi)$$

where the integration is done for each coefficient of the subsequent formal power series.

Proof. We have

$$\frac{\partial\Phi}{\partial t}(t,\tau)(x) = \tau.F(t)\circ\Phi(t,\tau)(x), \qquad \text{thus}$$

$$\frac{\partial}{\partial t}\frac{\partial\Phi}{\partial\tau}(t,\tau)(x) = F(t)\circ\Phi(t,\tau)(x) + \tau.TF(t)(\Phi(t,\tau)(x), \frac{\partial\Phi}{\partial\tau}(t,\tau)(x)) \quad .$$

This equation has a unique solution to any initial value, as the same arguments like in the proof of proposition 2 hold. On the other hand,

$$\frac{\partial}{\partial t}(T\Phi(t,\tau)(x,\int_0^t T\Phi(\xi,\tau)^{-1}(\Phi(\xi,\tau)(x),F(\xi)\circ\Phi(\xi,\tau)(x))d\xi) =$$

$$= \tau.TF(t)(\Phi(t,\tau)(x),T\Phi(t,\tau)(x,\int_0^t T\Phi(\xi,\tau)^{-1}(\Phi(\xi,\tau)(x),F(\xi)\circ\Phi(\xi,\tau)(x))d\xi) +$$

$$+ F(t)\circ\Phi(t,\tau)(x) \; .$$

Thus equation (**) holds, because $\Phi(0,\tau)(x) = x$ and therefore

$$\frac{\partial\Phi}{\partial\tau}(0,\tau)(x) = 0 = T\Phi(0,\tau)(x,0) \; .$$

Corollary. Using equation (**) in the definition of H, we get

$$H(t,\tau)(x) = \int_0^t T\Phi(\xi,\tau)^{-1}(\Phi(\xi,\tau)(x),F(\xi)\circ\Phi(\xi,\tau)(x))d\xi \quad . \qquad (^{***})$$

The following equations in proposition 4 are easily obtained from the identity

$$\Phi(t,\tau)\circ\Phi(t,\tau)^{-1}(x) = \Phi(t,\tau)^{-1}\circ\Phi(t,\tau)(x) = x$$

or from equation (***) by use of the chain rule and differentiation:

Proposition 4.

$$\frac{\partial}{\partial t}\,\tilde{\phi}(t,\tau)^{-1}(x) = -\tau.T\tilde{\phi}(t,\tau)^{-1}(x,F(t)(x))$$

$$\frac{\partial}{\partial \tau}\,\tilde{\phi}(t,\tau)^{-1}(x) = -H(t,\tau)\circ\tilde{\phi}(t,\tau)^{-1}(x)$$

$$\frac{\partial H}{\partial \tau}(t,\tau)(x) = \int_0^t T^2\tilde{\phi}(\xi,\tau)^{-1}(\tilde{\phi}(\xi,\tau)(x),T\tilde{\phi}(\xi,\tau)(x,H(\xi,\tau)(x)),F(\xi)\circ\tilde{\phi}(\xi,\tau)(x))d\xi\,+$$
$$+\int_0^t T\tilde{\phi}(\xi,\tau)^{-1}(\tilde{\phi}(\xi,\tau)(x),TF(\xi)(\tilde{\phi}(\xi,\tau)(x),T\tilde{\phi}(\xi,\tau)(x,H(\xi,\tau)(x))))d\xi\,-$$
$$-\int_0^t TH(\xi,\tau)(x,T\tilde{\phi}(\xi,\tau)^{-1}(\tilde{\phi}(\xi,\tau)(x),F(\xi)\circ\tilde{\phi}(\xi,\tau)(x)))d\xi\;.$$

The next lemma gives some very useful identities in the case that the original path ϕ has C^2-coefficients:

Proposition 5. Suppose that F has C^1-coefficients. Then the equations

(1) $$\frac{\partial}{\partial t}(T\tilde{\phi}(t,\tau)^{-1}(\tilde{\phi}(t,\tau)(x),F(t)\circ\tilde{\phi}(t,\tau)(x) =$$
$$= T\tilde{\phi}(t,\tau)^{-1}(\tilde{\phi}(t,\tau)(x),F'(t)\circ\tilde{\phi}(t,\tau)(x)$$

(2) $$\frac{\partial}{\partial t}(T\tilde{\phi}(t,\tau)^{-1}(\tilde{\phi}(t,\tau)(x),TF(t)(\tilde{\phi}(t,\tau)(x),T\tilde{\phi}(t,\tau)(x,y)))) =$$
$$= T\tilde{\phi}(t,\tau)^{-1}(\tilde{\phi}(t,\tau)(x),TF'(t)(\tilde{\phi}(t,\tau)(x),T(\tilde{\phi}(t,\tau)(x,y)))\,+$$
$$+\,\tau.T\tilde{\phi}(t,\tau)^{-1}(\tilde{\phi}(t,\tau)(x),T^2F(t)(\tilde{\phi}(t,\tau)(x);F(t)\circ\tilde{\phi}(t,\tau)(x),T\tilde{\phi}(t,\tau)(x,y)))$$

(3) $$\frac{\partial}{\partial t}(T^2\tilde{\phi}(t,\tau)^{-1}(\tilde{\phi}(t,\tau)(x);T\tilde{\phi}(t,\tau)(x,y),F(t)\circ\tilde{\phi}(t,\tau)(x))) =$$
$$= T^2\tilde{\phi}(t,\tau)^{-1}(\tilde{\phi}(t,\tau)(x);T\tilde{\phi}(t,\tau)(x,y),F'(t)\circ\tilde{\phi}(t,\tau)(x))\,-$$
$$-\,\tau.T\tilde{\phi}(t,\tau)^{-1}(\tilde{\phi}(t,\tau)(x),T^2F(t)(\tilde{\phi}(t,\tau)(x);T\tilde{\phi}(t,\tau)(x,y),F(t)\circ\tilde{\phi}(t,\tau)(x)))$$

hold. The proof is done by straightforward computation using the chain rule and the first formula stated in proposition 4.

Next we give a formula for the derivative of the pilgerschritt transform $\tilde{F}$ of F:

Proposition 6. Suppose that F has C^1-coefficients. Then the equation

$$\frac{\partial\hat{F}}{\partial\tau}(t,\tau)(x) =$$

$$= \int_0^t \int_0^\xi T\Phi(t,\tau)(\Phi(t,\tau)^{-1}(x), \int_\eta^\xi V(\xi,\eta,\mu) \circ \Phi(t,\tau)^{-1}(x)d\mu)d\eta d\xi$$

holds, where V denotes the FPS

$$V(\xi,\eta,\mu)(x) =$$

$$T\Phi(\mu,\tau)^{-1}(\Phi(\mu,\tau)(x), TF'(\mu)(\Phi(\mu,\tau)(x), T\Phi(\mu,\tau)(x,$$

$$T\Phi(\eta,\tau)^{-1}(\Phi(\eta,\tau)(x), F(\eta) \circ \Phi(\eta,\tau)(x))))) -$$

$$- T\Phi(\eta,\tau)^{-1}(\Phi(\eta,\tau)(x), TF(\eta)(\Phi(\eta,\tau)(x), T\Phi(\eta,\tau)(x,$$

$$T\Phi(\mu,\tau)^{-1}(\Phi(\mu,\tau)(x), F'(\mu) \circ \Phi(\mu,\tau)(x))))) +$$

$$+ T^2\Phi(\mu,\tau)^{-1}(\Phi(\mu,\tau)(x); T\Phi(\mu,\tau)(x, T\Phi(\eta,\tau)^{-1}(\Phi(\eta,\tau)(x),$$

$$F(\eta) \circ \Phi(\eta,\tau)(x))), F'(\mu) \circ \Phi(\mu,\tau)(x)) -$$

$$- T^2\Phi(\eta,\tau)^{-1}(\Phi(\eta,\tau)(x); T\Phi(\eta,\tau)(x, T\Phi(\mu,\tau)^{-1}(\Phi(\mu,\tau)(x),$$

$$F'(\mu) \circ \Phi(\mu,\tau)(x))), F(\eta) \circ \Phi(\eta,\tau)(x)) .$$

Proof. Using the result of the corollary to proposition 3 we get

$$\frac{\partial\hat{F}}{\partial\tau}(t,\tau)(x) =$$

$$= \frac{\partial}{\partial\tau}(T\Phi(t,\tau)(x, H(t,\tau) \circ \Phi(t,\tau)(x))) =$$

$$= \int_0^t \int_0^\xi T\Phi(t,\tau)(\Phi(t,\tau)^{-1}(x), W(\xi,\eta,\tau) \circ \Phi(t,\tau)^{-1}(x))d\eta d\xi, \text{ where}$$

$$W(\xi,\eta,\tau)(x) =$$

$$= T\Phi(\xi,\tau)^{-1}(\Phi(\xi,\tau)(x), TF(\xi)(\Phi(\xi,\tau)(x), T\Phi(\xi,\tau)(x,$$

$$T\Phi(\eta,\tau)^{-1}(\Phi(\eta,\tau)(x), F(\eta) \circ \Phi(\eta,\tau)(x))))) -$$

$$- T\tilde{\phi}(\eta,\tau)^{-1}(\tilde{\phi}(\eta,\tau)(x),TF(\eta)(\tilde{\phi}(\eta,\tau)(x),T\tilde{\phi}(\eta,\tau)(x,$$

$$T\tilde{\phi}(\xi,\tau)^{-1}(\tilde{\phi}(\xi,\tau)(x),F(\xi) \circ \tilde{\phi}(\xi,\tau)(x))))) +$$

$$+ T^2\tilde{\phi}(\xi,\tau)^{-1}(\tilde{\phi}(\xi,\tau)(x);T\tilde{\phi}(\xi,\tau)(x,T\tilde{\phi}(\eta,\tau)^{-1}(\tilde{\phi}(\eta,\tau)(x) ,$$

$$F(\eta) \circ \tilde{\phi}(\eta,\tau)(x))),F(\xi) \circ \tilde{\phi}(\xi,\tau)(x)) -$$

$$- T^2\tilde{\phi}(\eta,\tau)^{-1}(\tilde{\phi}(\eta,\tau)(x);T\tilde{\phi}(\eta,\tau)(x,T\tilde{\phi}(\xi,\tau)^{-1}(\tilde{\phi}(\xi,\tau)(x) ,$$

$$F(\xi) \circ \tilde{\phi}(\xi,\tau)(x))),F(\eta) \circ \tilde{\phi}(\eta,\tau)(x))$$

which leads directly to the desired result by use of the formulas stated in proposition 5.

The following lemma will be very useful in the subsequent theorem on pilgerschritt transform:

Proposition 7. Let ϕ, F, $\tilde{\phi}$, $\tilde{F}$ be as in definition 1. Then we have

(1) ϕ is a one-parameter-subgroup in G iff all the F_i are constant.

(2) If $F_1, F_2,\dots,F_n$ are constant, then

$$\tilde{F}_1 = F_1, \tilde{F}_2 = F_2,\dots,\tilde{F}_n = F_n$$

(3) $\tilde{F}_k(0) = \int_0^1 F_k(t)dt.$

Proof.

(1) $\phi_n(t+h) - \phi_n(t) = (\phi(h) \circ \phi(t))_n - \phi_n(t) =$

$$= \sum_{1\le k\le n} \sum_{i_1+\dots+i_k=n} \phi_k(h) \circ (\phi_{i_1}(t),\dots,\phi_{i_k}(t)) - \phi_n(t) =$$

$$= \sum_{1\le k\le n} \sum_{i_1+\dots+i_k=n} (\phi_k(h) - \phi_k(0)) \circ (\phi_{i_1}(t),\dots,\phi_{i_k}(t))$$

because $\phi(0)(x) = x$.

Thus $\phi'(t) = \phi'(0) \circ \phi(t)$, and $F = \phi'(0)$ is constant. On the other hand, for fixed s the FPS $\phi(t+s)$ and $\phi(t) \circ \phi(s)$ are both solutions of the differential equation $\psi'(t) = F \circ \psi(t)$ for the initial value $\phi(s)$ - thus these two series are identical.

(2) For computing the k-th homogeneous polynomial of a composite FPS only the first k coefficients of each of the FPS are necessary; the same is valid for computing the inverse of a FPS or Φ from F. As $F_1' = 0,\dots,F_n' = 0$, the formula of proposition 6 gives immediately $\tilde{F}_1' = 0,\dots,\tilde{F}_n' = 0$. Formula (2) is an immediate consequence of (3).

(3) As $\tilde{\phi}(0) = x$, we have

$$\tilde{F}(0) = \frac{\partial\tilde{\phi}}{\partial\tau}(0) = \int_0^1 F(\xi)(x)d\xi \qquad \text{by proposition 3.}$$

Theorem 3. Let F have C^2-coefficients, and use the notation as in definition 1.

(1) If $F_k'(t) = 0$ for all t, $1 \leq k \leq n$, then

$$\sup_{\tau\in[0,1]} \|\tilde{F}'_{n+1}(\tau)\| \leq$$

$$\leq \sup_{t\in[0,1]} \|F'_{n+1}(t)\| . \|F_1\| . \frac{(\|F_1\|(n+2)-2).e^{(n+2)\|F_1\|}+(n+2)\|F_1\|+2}{(n+2)^2\|F_1\|^3} .$$

(2) If F_k = const. for $1 \leq k \leq n$ and $F_1 = 0$, then we have $\tilde{F}'_{n+1} = 0$.

Proof. We use the FPS V of proposition 6: As $F_k'(t) = 0$ for $1 \leq k \leq n$, $V(\xi,\eta,\mu)_k = 0$ for $1 \leq k \leq n$, and the formula of proposition 6 reduces to

$$(\frac{\partial\hat{F}}{\partial\tau}(t,\tau))_{n+1}(x) =$$

$$= \int_0^t\int_0^\xi \Phi(t,\tau)_1 . \int_\eta^\xi V(\xi,\eta,\mu)_{n+1} \circ \Phi(t,\tau)_1^{-1}(x)d\mu d\eta d\xi =$$

$$= \int_0^t\int_0^\xi e^{t\tau F}1. \int_\eta^\xi e^{-\mu\tau F}1.(n+1).F'_{n+1}(\mu)(e^{\mu\tau F}1.e^{-t\tau F}1.x,\dots,$$

$$e^{\mu\tau F}1.e^{-\eta\tau F}1.F_1.e^{\eta\tau F}1.e^{-t\tau F}1.x)d\mu d\eta d\xi -$$

$$- \int_0^t\int_0^\xi e^{t\tau F}1. \int_\eta^\xi e^{-\eta\tau F}1.F_1.e^{\eta\tau F}1.e^{-\mu\tau F}1.F'_{n+1}(\mu)(e^{\mu\tau F}1.e^{-t\tau F}1.x,\dots,$$

$$e^{\mu\tau F}1.e^{-t\tau F}1.x)d\mu d\eta d\xi \quad .$$

As $e^{-\eta\tau F}1.F_1.e^{\eta\tau F}1 = F_1$, we have

$$\|\tilde{F}'(\tau)_{n+1}\| \leq$$

$$\leq \int_0^1 \int_0^\xi \int_\eta^\xi (n+2).\|F_1\|.\|F'_{n+1}(\mu)\|.e^{(n+2)(1-\mu)\tau\|F_1\|} d\mu d\eta d\xi \leq$$

$$\leq \sup_{t \in [0,1]} \|F'_{n+1}(t)\|.\|F_1\|.(n+2).\int_0^1 \int_0^\xi \int_\eta^\xi e^{(n+2)(1-\mu)\tau\|F_1\|} d\mu d\eta d\xi \quad .$$

Integration leads to the formula given above.

(2) This is an immediate consequence of (1), because $F_1 = 0$ in this case.

Corollary 1. Let $\phi(t)_1(x) = x$ for all t. Then we have $F_1 = 0$, and the sequence ϕ, $\tilde{\phi}$, $\tilde{\tilde{\phi}}$,... computes all the coefficients of a one-parameter-subgroup through the element $\phi(1)(x)$. The first n coefficients of the n-th element of this sequence agree with the first n coefficients of the one-parameter-subgroup.

Corollary 2. If $F'_k = 0$ for $1 \leq k \leq n$ and

$$\|F_1\|.\frac{(\|F_1\|(n+2)-2).e^{(n+2)\|F_1\|}+(n+2)\|F_1\|+2}{(n+2)^2\|F_1\|^3} < 1$$

then the (n+1)-th coefficient of the sequence ϕ, $\tilde{\phi}$, $\tilde{\tilde{\phi}}$,... converges uniformly to the (n+1)-th coefficient of a one-parameter subgroup through $\phi(1)(x)$.

Remark. The results in this paper do not depend on the dimension of the Lie group G (in the first part) respectively on the number of independents for the results on formal power series. As the proofs show, all the theorems and propositions are valid for Banach-Lie-groups respectively formal power series over Banach spaces, too.

REFERENCES

[1] Abraham, R., Marsden, J.E.: Foundations of mechanics. Benjamin Publ. Comp. 1978.

[2] Förg-Rob, W., Netzer, N.: Eine Methode zur Berechnung von einparametrigen Untergruppen ohne Verwendung des Logarithmus. Sitzungsberichte der Österreichischen Akademie der Wissenschaften, p.273-284, Wien 1981.

[3] Förg-Rob, W., Netzer, N.: Product integration and one-parameter subgroups of linear Lie groups. In this volume.

[4] Liedl, R., Netzer, N., Reitberger, H.: Über eine Methode zur Auffindung stetiger Iterationen in Lie-Gruppen. Aequationes Mathematicae 24, 19-32 (1982).

Wolfgang Förg-Rob
Institut für Mathematik
Universität Innsbruck
A-6020 Innsbruck

PRODUCT-INTEGRATION AND ONE-PARAMETER SUBGROUPS OF LINEAR LIE-GROUPS

Wolfgang Förg-Rob and Norbert Netzer*

ABSTRACT. The pilgerschritt transform is an iterative method which can be used to calculate one-parameter subgroups of groups of matrices, whose restrictions to the interval [0,1] belongs to a given homotopy class. The restrictions of one-parameter subgroups are fixed points of this method. All fixed points can be described by a product integral equation. Until now there could be shown for special groups of matrices that these fixed points are attractive. This paper deals with the group Aff(1,$\mathbb{R}$), where the product integral equation can be reduced to a Fredholm integral equation of the second kind. In this case the existence of one-parameter subgroups is proved, whose restrictions are not attractive fixed points. Furthermore, the existence of other fixed points is shown.

INTRODUCTION

This paper deals with a method using the theory of product integration (cf.Dollard-Friedman [1,2,3]), Schlesinger [13], Hostinsky [8]) to solve the following problem: Let G be a closed subgroup of the group of invertible matrices $Gl_n(\mathbb{R})$, $B \in G$ and $\phi:[0,1] \to G$ a C^1-path with the property $\phi(0) = E$(= unit matrix) and $\phi(1) = B$. We want to find a one-parameter subgroup $h:\mathbb{R} \to G$ such that $g: = h|[0,1]$ is homotopic to ϕ, or equivalently, we want to solve the boundary-value problem

$$Y''(t) = Y'(t)Y(t)^{-1}Y'(t)$$

$$Y(0) = E, \ Y(1) = B \tag{1}$$

under the restriction that the solution Y is homotopic to ϕ. For if there exists the one-parameter subgroup h, there is an element C of the Lie-algebra L(G) of G such that $h(t) = \exp(tC)$. Thus, we have $h'(t) = C.h(t)$ and $h''(t) = h'(t)h(t)^{-1}h'(t)$. On the other hand, every solution of (1) is a one-parameter subgroup.

In order to show some relation between the problem stated above and product integration we give a short view on product integration: Let $A:[a,b] \to L(G)$ be a continuous path. For $s \in [a,b]$ the product integral $\prod_a^s \exp(A(t)dt)$ is defined by the following process:
For a subdivision $\pi:a = t_0 < t_1 < \dots < t_{m+1} = s$ of the interval [a,s] set $\delta_k: = t_{k+1} - t_k$, $\theta_k \in [t_k,t_{k+1}]$, $|\pi|: = \max_{0\leq k\leq m} \delta_k$. Then the product integral is given by

*Presented by N. Netzer

$$\prod_a^s \exp(A(t)dt) :=$$

$$:= \lim_{|\pi|\to 0} \exp(\delta_m A(\theta_m)).\exp(\delta_{m-1}A(\theta_{m-1})). \ldots .\exp(\delta_o A(\theta_o)) =$$

$$= \lim_{|\pi|\to 0} (E+\delta_m A(\theta_m)).(E+\delta_{m-1}A(\theta_{m-1})). \ldots .(E+\delta_o A(\theta_o)) =:$$

$$=: \int_a^s (E+A(t)dt) \quad .$$

Considering this integral with s variable it can be shown that the function $s \to \prod_a^s \exp(A(t)dt)$ is the unique solution of the classical integral equation

$$Y(t) = E+\int_a^t A(\xi)Y(\xi)d\xi \quad .$$

Now let h be a one-parameter subgroup and $g = h|[0,1]$.

a) For each $\tau \in [0,1]$ we have

$$g(\tau) = \exp(\tau C) = \prod_0^1 \exp(\tau C dt) = \prod_0^1 \exp(\tau g'(t)g(t)^{-1}dt), \text{ i.e.,}$$

g is a solution of the product integral equation

$$\gamma(\tau) = \prod_0^1 \exp(\tau\gamma'(t)\gamma(t)^{-1}dt) \tag{2}$$

(cf. Dollard-Friedman [2], form. 108, 109).

b) For solving equation (2) we define inductively a sequence of C^1-functions by use of the product integral, starting with the path ϕ. Under suitable conditions this sequence will converge to the function g.

We define a new path $\tilde{\phi}:[0,1] \to G$, setting

$$\tilde{\phi}(\tau) := \prod_0^1 \exp(\tau\phi'(t)\phi(t)^{-1}dt)$$

The function $\tilde{\phi}$ has the following properties:

(1) If $\hat{\phi}:[0,1] \times [0,1] \to G$ is defined by

$$\hat{\phi}(t,\tau) := \prod_0^t \exp(\tau\phi'(\xi)\phi(\xi)^{-1}d\xi),$$

which means that $\hat{\Phi}$ is the solution of the classical integral equation

$$\hat{\Phi}(t,\tau) = E + \int_0^t \phi'(\xi)\phi(\xi)^{-1}\hat{\Phi}(\xi,\tau)d\xi \ , \qquad (3)$$

then we have $\tilde{\phi}(t) = \hat{\Phi}(1,t)$ for each $t \in [0,1]$.

(2) $\tilde{\phi}$ is an analytic function

(3) $\tilde{\phi}(0) = \phi(0) = E$

(4) $\tilde{\phi}(1) = \phi(1)$

(5) If $\pi: 0 = t_0 < t_1 < \dots < t_{m+1} = 1$ is a subdivision of the interval [0,1] and δ_k, $|\pi|$ are defined as above, we have

$$\tilde{\phi}(\tau) = \lim_{|\pi|\to 0} \phi(t_m+\tau\delta_m)\phi(t_m)^{-1} \dots \phi(t_0+\tau\delta_0)\phi(t_0)^{-1} .$$

(cf. Liedl [10])

(6) If ϕ is the restriction of a one-parameter subgroup to the interval [0,1], we have $\tilde{\phi} = \phi$.

Now we define inductively the sequence $(\overset{n}{\tilde{\phi}})_{n \in N}$ by

$$\overset{o}{\tilde{\phi}} := \phi, \qquad \overset{n+1}{\tilde{\phi}} := \tilde{\overset{n}{\tilde{\phi}}} \ .$$

This sequence will be called the sequence of pilgerschritt transforms of ϕ or pilgerschritt sequence of ϕ. It is easy to see that this method is analogous to iterative methods for finding solutions of classical integral equations. In the papers [4], [10] and [12] sufficient conditions are given for the uniform convergence of the pilgerschritt sequence of ϕ to a solution γ of equation (2). Liedl [10] has shown that ϕ and γ are homotopic in this case. In the general case (i.e., if there are no conditions on the structure of the group G) the uniform convergence of the pilgerschritt sequence has been proved under the assumption that the image of ϕ lies in a sufficiently small neighbourhood of E and that $\|\phi'(t)\|$ is small enough for $t \in [0,1]$ ([4]). By studying several groups admitting special structures, the uniform convergence of the pilgerschritt sequence can be proved for a larger class of functions ϕ:

If G is a commutative group, $\tilde{\phi}$ is a solution of equ.(2) for every C^1-path ϕ (cf.[10]).

If G is nilpotent of degree n, $\overset{n}{\tilde{\phi}}$ is a solution of equ.(2) for every C^1-path ϕ (cf.[12]).

If G is a group of triangular matrices and $\phi:[0,1] \to G$ is C^1, the pilgerschritt sequence of ϕ converges uniformly to a solution of equ.(2), whenever the elements of the diagonal of $\phi(t)$ do not differ too much from 1, for all $t \in [0,1]$ (cf.[9]). In each of the cases stated above the pilgerschritt sequence leads to restrictions $g = h|[0,1]$ of one-parameter subgroups $h: \mathbb{R} \to G$. If in these cases g is perturbed to a path ϕ such that $\|\phi'(t) - g'(t)\|$ is small enough for all $t \in [0,1]$, the sequence of pilgerschritt transforms of ϕ converges uniformly to g. This fact gives rise to the following questions:

a) Is the result stated above true for every one-parameter subgroup?

b) Are there any solutions of the equation (2) which are not restrictions of one-parameter subgroups?

Specializing G to be the connected component of the identity of the group $Aff(1,\mathbb{R})$, that is the group

$$\left\{\begin{pmatrix} 1 & y \\ 0 & e^x \end{pmatrix} \middle| x,y \in \mathbb{R}\right\},$$

equation (2) can be reduced to a Fredholm integral equation of the second kind. Thus by the classical theory of these equations one can prove that the first question has to be answered by "no" and the second by "yes". Furthermore, we prove that the pilgerschritt sequence is oscillating for special functions ϕ, that is, we have $\tilde{\phi} \neq \phi$, $\overset{2n}{\tilde{\phi}} = \phi$ and $\overset{2n+1}{\tilde{\phi}} = \tilde{\phi}$ for all $n \in \mathbb{N}$.

1. On the convergence of the pilgerschritt sequence

For each $\begin{pmatrix} 1 & y \\ 0 & e^x \end{pmatrix} \in G$ there exists a uniquely determined element $C \in L(G)$ satisfying $\exp(C) = \begin{pmatrix} 1 & y \\ 0 & e^x \end{pmatrix}$, namely

$$\begin{pmatrix} 0 & y\dfrac{x}{e^x-1} \\ 0 & x \end{pmatrix} .$$

Therefore every path $\phi:[0,1] \to G$ can be decomposed in one and only one way by

$$\phi(t) = \exp(t.\begin{pmatrix} 0 & b \\ 0 & a \end{pmatrix}).\begin{pmatrix} 1 & \beta(t) \\ 0 & \alpha(t) \end{pmatrix} ,$$

where α and β denote real valued functions satisfying $\alpha(0) = \alpha(1) = 1$ and $\beta(0) = \beta(1) = 0$. Then we have

$$\phi(1) = \exp\left(\begin{pmatrix} 0 & b \\ 0 & a \end{pmatrix}\right) \text{ and } \tilde{\phi}(t) = \exp\left(t.\begin{pmatrix} 0 & b \\ 0 & a \end{pmatrix}\right).\begin{pmatrix} 1 & \tilde{\beta}(t) \\ 0 & 1 \end{pmatrix} \quad \text{for all}$$

$t \in [0,1]$, where $\tilde{\beta}:[0,1] \to \mathbb{R}$ is a C^∞-function, and $\tilde{\beta}(0) = \tilde{\beta}(1) = 0$. Therefore we may assume $\alpha(t) = 1$ for each $t \in [0,1]$ without loss of generality.

<u>Proposition.</u> The pilgerschritt transform of the path

$$\phi:[0,1] \to G:t \to \exp\left(t.\begin{pmatrix} 0 & b \\ 0 & a \end{pmatrix}\right).\begin{pmatrix} 1 & \beta(t) \\ 0 & 1 \end{pmatrix}, \ \beta(0) = \beta(1) = 0, \quad \text{is}$$

given by $\tilde{\phi}(t) = \exp\left(t.\begin{pmatrix} 0 & b \\ 0 & a \end{pmatrix}\right).\begin{pmatrix} 1 & \tilde{\beta}(t) \\ 0 & 1 \end{pmatrix}$, where $\tilde{\beta}:[0,1] \to \mathbb{R}$ is

the function defined by $\tilde{\beta}(\tau) = (1-\tau)\tau a\int_0^1 e^{(\tau-1)ta}\beta(t)dt$.

<u>Proof.</u> The function

$$Y(\tau,t) = \exp\left(t\tau.\begin{pmatrix} 0 & b \\ 0 & a \end{pmatrix}\right).\begin{pmatrix} 1 & \tau\int_0^t \beta'(\xi)e^{(\tau-1)\xi a}d\xi \\ & 1 \end{pmatrix}$$

is the unique solution of the classical integral equation (3). So we have

$\tilde{\beta}(\tau) = \tau\int_0^1 \beta'(t)e^{(\tau-1)ta}dt$. Integration by parts leads to

$\tilde{\beta}(\tau) = \tau(\beta(1)e^{(\tau-1)a} - \beta(0)) - (\tau-1)\tau a\int_0^1 e^{(\tau-1)ta}\beta(t)dt$.

Since $\beta(0) = \beta(1) = 0$, the proposition is proved. This result gives rise to a first theorem on convergence respectively divergence of the pilgerschritt sequence:

<u>Theorem.</u> (1) If $a > \ln(\frac{1}{2})$, the sequence $(\overset{n}{\phi})_{n \in N}$ converges uniformly to the one-parameter subgroup

$$t \to \exp\left(t.\begin{pmatrix} 0 & b \\ 0 & a \end{pmatrix}\right).$$

(2) Suppose $c,d \in \mathbb{R}$ such that $0 < c < d < 1$. For a sufficiently small and for every $\varepsilon > 0$ there exists a path ϕ such that $\phi(1) = \exp\left(\begin{pmatrix} 0 & b \\ 0 & a \end{pmatrix}\right)$, $\sup_{t\in[0,1]} |\beta(t)| \leq \varepsilon$ and the sequence $(\overset{n}{\phi}(t))_{n \in N}$ diverges for each $t \in [c,d]$.

Proof. (1) Let $M := \sup_{t \in [0,1]} |\beta(t)|$. For $a > 0$ we have

$$|\tilde{\beta}(\tau)| \le M \cdot \int_0^1 (1-\tau) a e^{t(\tau-1)a} dt = M \cdot (1-e^{(\tau-1)a}) \le M \cdot (1-e^{-a}) .$$

Since $1-e^{-a} < 1$, (1) is proved for positive numbers a. If $a = 0$, we have $\tilde{\beta}(t) = 0$ for each $t \in [0,1]$.

If $a < 0$, we have $|\beta(\tau)| \le M.\int_0^1 (\tau-1) a e^{t(\tau-1)a} dt = M.(e^{(\tau-1)a}-1) \le M(e^{-a}-1)$. As $e^{-a}-1 < 1$ is equivalent to $a > \ln(\frac{1}{2})$, part (1) of the theorem is proved.

(2) Given c,d such that $0 < c < d < 1$ we choose $a < 0$ such that $L := c.e^{(d-1)ac}(e^{(d-1)a(d-c)}-1) > 1$. Furthermore, choose a function $\beta: [0,1] \to \mathbb{R}$ having the properties $\beta(t) \ge 0$ for all $t \in [0,1]$ and $\frac{\varepsilon}{2} \le \beta(t) \le \varepsilon$ for each $t \in [c,d]$. Then we have

$\tilde{\beta}(\tau) \le 0$ for all $\tau \in [0,1]$ and

$$\tilde{\beta}(\tau) \le -\tau \int_c^d (\tau-1) a \beta(t) e^{(\tau-1)ta} dt \le -\frac{\varepsilon}{2}\tau \int_c^d (\tau-1) a e^{(\tau-1)ta} dt =$$

$$= -\frac{\varepsilon}{2}\tau (e^{(\tau-1)ad} - e^{(\tau-1)ac}) \le$$

$$\le -\frac{\varepsilon}{2} c e^{(d-1)ac}(e^{(d-1)a(d-c)} - 1) \le -L\frac{\varepsilon}{2}$$

for each $\tau \in [c,d]$. Analogously it can be shown that $\tilde{\tilde{\beta}}(\tau) > 0$ for all $\tau \in [0,1]$ and $\tilde{\tilde{\beta}}(\tau) \ge L^2 \frac{\varepsilon}{2}$ for each $\tau \in [c,d]$. It is easy to see that

$$|\overset{n}{\tilde{\beta}}(\tau)| \ge L^n \cdot \frac{\varepsilon}{2} \quad \text{for each } \tau \in [c,d].$$

For getting further results we set

$$K_a(\tau,t) := (1-\tau)\, a e^{(\tau-1)ta}$$

and study the action of the integral operator K_a defined by the kernel $K_a(\tau,t)$ on the Hilbert space $L_2([0,1])$ having the usual inner product

$$\langle f,g \rangle := \int_0^1 f(t)\overline{g(t)} dt \quad .$$

So $K_a f$ is defined by

$$(K_a f)(\tau) := \int_0^1 K_a(\tau,t)f(t)dt$$

for each $f \in L_2([0,1])$. (cf.[5]). Studies on the pilgerschritt transform of a path having the endpoint $\exp(\begin{pmatrix} 0 & b \\ 0 & a \end{pmatrix})$ are thus reduced to studies on the operator K_a. According to the notation in [5] we call a complex number $\lambda \neq 0$ an eigengalue of K_a iff the Fredholm equation of the second kind

$$y(\tau) = \frac{1}{\lambda} \int_0^1 K_a(\tau,t)y(t)dt$$

has a nontrivial solution. If $\phi \neq 0$ is a solution of this equation, i.e., if $K_a\phi = \lambda\phi$, ϕ will be called an eigenfunction belonging to λ. For fixed $a \in \mathbb{R}$ we write $K(\tau,t)$ instead of $K_a(\tau,t)$. Now we can state the following

Theorem. The sequence of iterated pilgerschritt transforms converges uniformly to the one-parameter subgroup $\exp(t.\begin{pmatrix} 0 & b \\ 0 & a \end{pmatrix})$ for every C^1-path in G joining the points $\begin{pmatrix} 1 & 0 \\ 0 & 1 \end{pmatrix}$ and $\exp(\begin{pmatrix} 0 & b \\ 0 & a \end{pmatrix})$ iff each eigenvalue λ of the operator K has the property $|\lambda| < 1$.

Proof.(1) Instead of the kernel $K(\tau,t) = a\tau(1-\tau)e^{-at(1-\tau)}$ we study the symmetric kernel

$$H_a(\tau,t) := a\sqrt{\tau(1-\tau)}.e^{-\frac{a}{2}\tau}e^{at\tau}e^{-\frac{a}{2}t}.\sqrt{t(1-t)}$$

(cf.[7]), which will be denoted by $H(\tau,t)$ for fixed a. It is well known that a function $\psi:[0,1] \to C$ is an eigenfunction of H belonging to λ iff

$$\phi:[0,1] \to C: t \to \sqrt{t(1-t)}e^{\frac{a}{2}t}\psi(t)$$

is an eigenfunction of the kernel K belongint to λ.

For the following assertions we refer the reader to [5]. As $H(\tau,t)$ is a symmetric kernel, it has only real eigenvalues. The set of eigenvalues is nonempty, bounded and has no limit points except possibly 0. There exists a sequence $(\psi_n)_{n \in N}$ of orthonormal eigenfunctions of H such that the series

$$\sum_{n \in N} < Hf,\psi_n > \psi_n$$

converges uniformly to Hf for all $f \in L_2([0,1])$.

Now we put

$$\beta^*(\tau): = \sqrt{\tau(1-\tau)}.e^{-\frac{a}{2}\tau}.\int_0^1 ae^{t(\tau-1)a}\beta(t)dt.$$

Then we have

$$(K^n\beta)(t) = \sqrt{t(1-t)}\, e^{\frac{a}{2}t}(H^{n-1}\beta^*)(t) =$$

$$= \sqrt{t(1-t)}e^{\frac{a}{2}t}.\sum_{m\in N} < H\beta^*,\psi_m > \lambda_m^{n-2}\psi_m(t) \quad \text{for} \quad n > 1, \quad \text{where } \lambda_m$$

denotes the eigenvalue belonging to ψ_m.

Now we set $M: = \sup_{m\in N}|\lambda_m| = \max_{m\in N}|\lambda_m|$. For $t \in [0,1]$ and $n > 1$ we have

$$|\overset{n}{\tilde{\beta}}(t)| = |(K^n\beta)(t)| \leq M^{n-2}|\sqrt{t(1-t)}e^{\frac{a}{2}t}\sum_{m\in N} < H\beta^*,\psi_m > \psi_m(t)| =$$

$$= M^{n-2}|\sqrt{t(1-t)}e^{\frac{a}{2}t}(H\beta^*)(t)| \leq$$

$$\leq M^{n-2}\sup_{\xi\in[0,1]}|\sqrt{\xi(1-\xi)}e^{\frac{a}{2}\xi}(H\beta^*)(\xi)| .$$

This inequality implies the uniform convergence of the pilgerschritt sequence to $\exp(t.\begin{pmatrix}0 & b\\ 0 & a\end{pmatrix})$ under the assumption $M < 1$.

(2) If $M \geq 1$, we choose an eigen-alue λ with the property $|\lambda|\geq 1$ and an eigenfunction χ belonging to λ. Now let $\phi(t): = \exp(t.\begin{pmatrix}0 & b\\ 0 & a\end{pmatrix}).\begin{pmatrix}1 & \chi(t)\\ 0 & 1\end{pmatrix}$. The following cases are of interest:

$\lambda = 1$: Then we have $\phi = \overset{n}{\tilde{\phi}}$ for all $n \in N$. Thus there exists a fixed point which is not a one-parameter subgroup.

$\lambda = -1$: Then we have $\phi = \overset{2n}{\tilde{\phi}}$, $\tilde{\phi} = \overset{2n+1}{\tilde{\phi}}$ and $\phi \neq \tilde{\phi}$ for all $n \in N$. Thus in this case the pilgerschritt sequence oscillates.

$|\lambda| > 1$: The sequence of pilgerschritt transforms diverges.

2. On the solutions of the product integral equation (2)

We show that there exist real numbers a_0 and a_1 such that -1 resp. 1 is an eigenvalue of K_{a_0} resp. K_{a_1}. For $a \in \mathbb{R}$ we set

$$\lambda_{max}(a) := \max\{\lambda \mid \lambda \text{ is an eigenvalue of } K_a\} \text{ and}$$

$$\lambda_{min}(a) := \min\{\lambda \mid \lambda \text{ is an eigenvalue of } K_a\}.$$

By a theorem of Courant (cf.[5],[13]) the equations

$$\lambda_{max}(a) = \sup \{\langle H_a f,f\rangle \mid \|f\|_2 = 1\} \text{ and}$$

$$\lambda_{min}(a) = \inf \{\langle H_a f,f\rangle \mid \|f\|_2 = 1\} \text{ are valid.}$$

It is easy to see that the functions $\lambda_{max}: R \to \mathbb{R}$ and $\lambda_{min}: R \to \mathbb{R}$ are continuous.

Theorem.

(1) There exists a real number a_0 such that $\lambda_{min}(a_0) = -1$.

(2) There exists a real number a_1 such that $\lambda_{max}(a_1) = +1$.

Proof. We show that there exist real numbers a_0' and a_1' such that $\lambda_{min}(a_0') \leq -1$ and $\lambda_{max}(a_1') \geq 1$. As all the eigenvalues have a modulus less than 1 for all $a > \ln(\frac{1}{2})$, the assertion is a trivial consequence of the intermediate-value-theorem.

(1) Let $\beta(t) = 1$ for all $t \in [0,1]$. Then we have

$$\langle H_a \beta,\beta\rangle = \frac{ae^{-\frac{a}{4}}}{16} \cdot \int_{-1}^{1}\int_{-1}^{1} \sqrt{1-y^2}\,\sqrt{1-x^2}\,e^{\frac{a}{4}xy}\,dxdy.$$

We may conclude for $a < 0$: $\lambda_{min}(a) \leqq \langle H_a \beta,\beta\rangle \leqq \frac{\pi^2}{64}a$. Now choose $a_0' \leqq -\frac{64}{\pi^2}$, and (1) is proved.

(2) Let $\beta(t) = \operatorname{sign}(t-\frac{1}{2})$ for $t \in [0,1]$. Then we have

$$\langle H_a \beta,\beta\rangle = \frac{ae^{-\frac{a}{4}}}{16} \cdot \Big(\int_{-1}^{0}\int_{-1}^{0} f(x,y)dxdy + \int_{0}^{1}\int_{0}^{1} f(x,y)dxdy -$$

$$- \int_{-1}^{0}\int_{0}^{0} f(x,y)dxdy - \int_{0}^{1}\int_{-1}^{0} f(x,y)dxdy\Big),$$

where $f(x,y) := \sqrt{1-x^2}\,\sqrt{1-y^2}\,e^{\frac{a}{4}xy}$.

For $a < 0$ we have

$$\int_{-1}^{0}\int_{-1}^{0} f(x,y)dxdy = \int_{0}^{1}\int_{0}^{1} f(x,y)dxdy \leqq \frac{\pi^2}{16} \quad \text{and}$$

$$\int_{-1}^{0}\int_{0}^{1} f(x,y)dxdy = \int_{0}^{1}\int_{-1}^{0} f(x,y)dxdy \geqq \frac{\pi^2}{16} + (e^{-\frac{a}{16}} - 1)c ,$$

where

$$c = \int_{\frac{1}{2}}^{1}\int_{-1}^{-\frac{1}{2}} \sqrt{1-x^2}\,\sqrt{1-y^2}\,dxdy > 0 .$$

Therefore we get

$$\lambda_{max}(a) \geqq \langle H_a\beta,\beta\rangle \geqq -a\frac{c}{4}e^{-\frac{a}{4}}(e^{-\frac{a}{16}} - 1) := h(a) .$$

As $\lim_{a\to-\infty} h(a) = \infty$, there exists a_1' such that $\lambda_{max}(a_1') \geqq 1$.

Remark. N. Ortner (Innsbruck) has shown that

$$\langle H_a\beta,\beta\rangle = \frac{\pi}{4}e^{-\frac{a}{4}}.(1-C-\log(\tfrac{a}{4}) - \tfrac{4}{a}.\mathrm{sh}(\tfrac{a}{4}) + \mathrm{Chi}(\tfrac{a}{4})) \quad \text{for} \quad \beta(t) = 1,$$

where C is the Euler-Mascheroni constant.

Remark. Suppose $a \in \mathbb{R}$. We decompose $L_2([0,1])$ into the direct sum $E_{1,a} \oplus E_{2,a} \oplus E_{3,a} \oplus \mathrm{Ker}(K_a)$, where $E_{1,a}$ denotes the subspace spanned by the eigenfunctions belonging to an eigenvalue λ with $|\lambda| > 1$ or $\lambda = -1$, $E_{2,a}$ is the eigenspace of the eigenvalue $\lambda = 1$, and $E_{3,a}$ is spanned by the eigenfunctions not contained in $E_{1,a} \oplus E_{2,a}$. The sequence $(K_a^n\phi)_{n \in N}$ converges uniformly for every $\phi \in L_2([0,1])$ iff $E_{1,a} = \{0\}$. If $a \geq 0$, then $|\lambda| < 1$ for each eigenvalue λ of K_a, that means $E_{1,a} \oplus E_{2,a} = \{0\}$. Therefore $E_{2,a} \neq \{0\}$ implies $a < 0$. In this case we have $H_a(\tau,t) \leq 0$ for $\tau,t \in [0,1]$, and thus the eigenvalue of H_a having the largest modulus - and of course that of K_a, too - must be negative (cf.[1]), so we also have $E_{1,a} \neq \{0\}$.Thus we see: If there exists a solution γ of the product integral equation (2) which is not a one-parameter subgroup having the property $\gamma(1) = \exp(\begin{pmatrix} 0 & 0 \\ 0 & a \end{pmatrix})$, in other words, if $E_{2,a} \neq \{0\}$, the pilgerschritt sequence of an arbitrarily chosen path $\phi:[0,1] \to G:t \to \exp(t\begin{pmatrix} 0 & 0 \\ 0 & a \end{pmatrix}).\begin{pmatrix} 1 & \beta(t) \\ 0 & 1 \end{pmatrix}$, $\beta(0) = \beta(1) = 0$, in general does not converge to the solution γ of (2), but diverges or oscillates. It is convergent to a solution iff $\beta = \beta_1 + \beta_2$, $0 \neq \beta_1 \in E_{2,a}$ and $\beta_2 \in E_{3,a} \oplus \mathrm{Ker}(K_a)$, and this solution is not the restriction of a one-parameter subgroup. The limit function is the restriction of a one-parameter subgroup iff $\beta \in E_{3,a} \oplus \mathrm{Ker}(K_a)$.

REFERENCES

[1] Dollard, J.D., and Friedman, C.N.: "Product Integration". Encyclopedia of mathematics and its applications, Addison-Wesley, London, 1979.

[2] Dollard, J.D., and Friedman, C.N.: On strong product integration. J. Func. Anal. 28, 309-354 (1978).

[3] Dollard, J.D., and Friedman, C.N.: Product integrals II: Contour integrals. J. Func. Anal. 28, 355-368 (1978).

[4] Förg-Rob, W., and Netzer, N.: Eine Methode zur Berechnung von einparametrigen Untergruppen ohne Verwendung des Logarithmus. Sitzungsberichte der Österreichischen Akademie der Wissenschaften II/190, 273-284 (1981).

[5] Gohberg, I.C., and Krein, M.G.: "Introduction to the Theory of Linear Nonselfadjoint Operators". Translations of Mathematical Monographs, vol.18, American Mathematical Society, Providence, R.I., 1969.

[6] Guggenheimer, H.W.: "Differential Geometry". Mac Graw-Hill, New York, 1963.

[7] Hoheisl, G.: "Integralgleichungen". Sammlung Göschen Band Nr. 1099, Berlin-Leipzig, 1936.

[8] Hostinsky, B., and Volterra, V.: "Operations Infinitesimales Lineaires", Paris, 1938.

[9] Kuhnert, K.: Die Konvergenz des Pilgerschrittverfahrens für unipotente und auflösbare lineare Gruppen. Berichte der mathematisch-statistischen Sektion im Forschungszentrum Graz Nr. 87 (1978).

[10] Liedl, R.: Über eine Methode zur Lösung der Translationsgleichung. Berichte der mathematisch-statistischen Sektion im Forschungszentrum Graz, Nr.84 (1978).

[11] Liedl, R., Netzer, N., Reitberger, H.: Über eine Methode zur Auffindung stetiger Iterationen in Lie-Gruppen, Aequationes Mathematicae 24,19-32 (1982).

[12] Netzer, N., and Reitberger, H.: On the convergence of iterated pilgerschritt transformations in nilpotent Lie Groups. Publicationes Mathematicae Debrecen 29, 309-314 (1982).

[13] Schlesinger, L.: Neue Grundlagen für einen Infinitesimalkalkül der Matrizen. Math. Zeit. 33, 33-61 (1931).

[14] Zabreyko, P.P. et al.: "Integral equations - a reference text". Noordhoff International Publishing Leyden, 1975.

W. Förg-Rob and N. Netzer
Institut für Mathematik
Univ. Innsbruck, Innrain 52
A-6020 Innsbruck

THE PERTURBATIVE METHOD FOR DISCRETE PROCESSES AND ITS PHYSICAL APPLICATION

Gêza Györgyi

ABSTRACT

In this note chaotic iterations are considered which only slightly differ from processes of known properties. Averages are calculated in case of noisy perturbation of systems exhibiting originally a fixed point and where the iteration goes along a discrete one-dimensional chain or on the Cayley tree. As a physical example the disordered Ising model is considered on these lattices and results on spin fluctuations are presented. Finally, probabilistic properties of perturbed one-dimensional iterations are reviewed, where even the unperturbed systems show chaotic behaviour. For the latter ones in the examples the hat function and the logistic map in the fully developed chaotic state are taken, and we also discuss a noisy iteration describing a one-dimensional disordered quantum mechanical system.

I. INTRODUCTION

Stochastic properties of chaotic systems constitute a subject of current interest. As for probability densities and certain averaged quantities of erratic processes an abundance of physical experiments and computer simulations is available (see e.g.[1-5]). On the other hand, calculations resulting in explicit formulae could be carried out so far mainly for some of the simplest chaotic discrete semi-dynamic processes. In the following we consider several examples, for which the corresponding probability densities and averaged quantities are evaluated by using perturbative methods. Such a technique might be applied if the system studied is not far from a one of known characteristics. As we shall see, however, there is no general receipt for perturbative calculations, especially when the unperturbed process is chaotic as well. In the selection of the following problems particular emphasis has been put on physical applications.

II. NOISY PERTURBATION OF ONE DIMENSIONAL ITERATIONS WITH A FIXED POINT

We first consider the one-dimensional iteration

$$x_{n+1} = f(x_n) + \varepsilon\zeta_n \quad , \tag{2.1}$$

where ε is a small parameter and $\zeta_1, \zeta_2 \ldots$ are uncorrelated random numbers distributed according to the symmetric density $\rho(\zeta)$. We assume an odd symmetry for the C^∞ function f with $0 < f'(0) < 1$ and $f'''(0) < 0$, the origin being a stable fixed point for $\varepsilon = 0$. The equation determining the stationary probability density P(x) of the iteration (2.1) reads as

$$P(x) = \int \rho(\rho) P(x') \delta(x - f(x') - \varepsilon\zeta) dx' d\zeta \tag{2.2}$$

with δ being the Dirac delta operation. The resulting probability density P(x) is symmetric again, and for the momenta of x one gets from Eq.(2.2)

$$\langle x^2 \rangle = \varepsilon^2 \langle \zeta^2 \rangle + \langle f^2(x) \rangle \quad , \tag{2.3a}$$

$$\langle x^4 \rangle = \varepsilon^4 \langle \zeta^4 \rangle + 6\varepsilon^2 \langle \zeta^2 \rangle \langle f^2(x) \rangle + \langle f^4(x) \rangle \quad , \tag{2.3b}$$

and so on, where <> means the average according to the corresponding density. The stability of the fixed point ensures the narrowness of P(x) for small x, so Eqs. (2.3) provide us with a systematic expansion for the momenta of x in the powers of ε. We quote the lowest order terms as

$$\langle x^2 \rangle = \varepsilon^2 \frac{\langle \zeta^2 \rangle}{1 - f'^2(0)} + \ldots \; , \tag{2.4a}$$

$$\langle x^4 \rangle = \varepsilon^4 \frac{1}{1 - f^4(0)} \{ \langle \zeta^4 \rangle + \frac{6f'^2(0)}{1 - f'^2(0)} \langle \zeta^2 \rangle^2 \} + \ldots, \tag{2.4b}$$

and so on, where... means higher order terms in ε. We mention that for a Gaussian density $\rho(\zeta)$ one has a Gaussian P(x) in the first approximation.

It is obvious that if the fixed point of the unperturbed map loses its stability, i.e. $\varepsilon^2 \gg (1 - f'^2(0)) \to 0$, then the expansions become divergent. We call this a critical situation, and expect that the leading term for any momentum $\langle x^{2n} \rangle$ contains now a smaller power of ε than in Eqs.(2.4). Unfortunately, for $f'(0) = 1$ Eqs.(2.3) do not furnish a systematic solution, so one has to adopt a different method. For this purpose the Kramers-Moyal [6,7] technique seems to be suitable.*)

This method consists in transforming Eq.(2.2) into a differential equation of infinite order as

$$P(x) = \sum_{n=0}^{\infty} \frac{(-1)^n}{n!} \frac{d^n}{dx^n} a_n(x)P(x) \quad , \tag{2.5}$$

where

$$a_0(x) = \int \rho(\zeta)(\varepsilon\zeta + f(x) - x)^n \, d\zeta \quad . \tag{2.6}$$

If one is interested in the lowest order term for the averages and the density $\rho(\zeta)$ is Gaussian then it is sufficient to retain only the first three terms on the right-hand-side of Eq.(2.5). The resulting differential equation can be solved as

$$P(x) \approx \frac{\text{const.}}{1+(x/a)^6} \exp \{- \frac{12}{f'''(0)a^{1/3}} \int_0^{x/a} \frac{z^3 dz}{1+z^6}\} \quad , \tag{2.7}$$

where $a = 36\varepsilon^2 <\zeta^2>/f'''^2(0)$. Hence the leading terms of the momenta $<x^{2n}>$ can be calculated for $\varepsilon \to 0$ by using the saddle-point method:

$$<x^{2n}> = \varepsilon^n \left[\frac{12<\zeta^2>}{f'''(0)}\right]^{\frac{n}{2}} \frac{\Gamma(\frac{2n+1}{4})}{\Gamma(\frac{1}{4})} + \ldots \quad . \tag{2.8}$$

As for higher order corrections, instead of using only (2.5) one rather applies the low order results of the Kramers-Moyal technique together with Eqs.(2.3), which simplifies the calculations considerably.

III. THE ONE-DIMENSIONAL ISING MODEL IN A DISORDERED EXTERNAL MAGNETIC FIELD

The Hamiltonian of the system reads as

$$- \frac{\mathcal{K}}{k_B T} = K \sum_{l=0}^{\infty} s_i s_i \quad + \sum_{i=0}^{\infty} h_i s_i \quad , \tag{3.1}$$

where the spin varialbe s_i associated with the site i may assume the values ± 1, k_B

+)At this point the author wishes to thank Professor H. Thomas for useful discussions.

is the Boltzmann factor, T denotes the temperature, $K = J/(k_b T) > 0$ represents the interaction between neighbouring spins, and h_i is the random magnetic field. The field is distributed on each site according to the density $\rho(h)$ and different site's fields are uncorrelated. It turns out [8-10] that the thermodynamics of the system can be described by determining the probability density P(x) of an iteration of the type (2.1) with

$$f(x) = \frac{1}{2} \ln \frac{\cosh(K+x)}{\cosh(K-x)}, \quad \varepsilon\zeta_n = h_{n+1} \quad . \tag{3.2}$$

In particular, the free energy per spin is

$$-\frac{F}{k_B T} = \frac{1}{2} \langle \ln[4 \cosh(K+x) \cosh(K-x)] \rangle \tag{3.3}$$

whereas the magnetization is zero and the spatial spin fluctuations can be characterized by

$$Q = \lim_{N\to\infty} \frac{1}{N} \sum_{i=0}^{N} \bar{s}_i^2 = \langle \tanh^2(x + f(x')) \rangle_{x,x'} \quad . \tag{3.4}$$

Here $\bar{s}_i$ denotes a thermal average of the spin on site i, i.e. the local magnetization at that site, and x,x' are independent random variables distributed according to P(x). We are now in a position to apply the results of Section II, which yield

$$-\frac{F}{k_B T} = \ln 2\cosh K + \frac{\langle h^2 \rangle}{2} - \frac{v^2}{2(1-v^2)} \langle h^2 \rangle^2 - \frac{\langle h^4 \rangle}{12} + \dots \quad , \tag{3.5}$$

$$Q = \frac{1+v^2}{1-v^2} \langle h^2 \rangle - \frac{4v^2(2+v^2)}{(1-v^2)^2} \langle h^2 \rangle^2 - \frac{2}{3} \frac{1+v^2}{1-v^2} \langle h^4 \rangle + \dots \, , \tag{3.6}$$

where the notation $v = \tanh K$ has been introduced. The expansion for F can also be obtained by evaluating products of random matrices [11]. For $v \to 1$, i.e. if the temperature approaches zero, no analytic expansions can be given for F and Q (see [12]).It is nevertheless apparent from Eq.(3.6) that the spin fluctuations along the chain become strong even for a small random magnetic field. Unfortunately, due to the non differentiable shape of f for $v = 1$, the critical fluctuations in the region $\langle h^2 \rangle \gg (1-v^2)$ cannot be described by the Kramers-Moyal method.

IV. NOISY PERTURBATION OF ITERATIONS ON THE CAYLEY TREE - THE RANDOM EXCHANGE ISING MODEL IN A DISORDERED FIELD

The iteration (2.1) can be generalized in several ways. Firstly, we allow parametric noise in addition to ζ. This occurs by taking $f(v,x)$ with a v distributed according to a given density $\omega(v)$, not necessarily a sharp function, and the numbers at different steps are uncorrelated. Secondly, instead of going along a one dimensional chain, we advance the iteration on a sort of the Cayley tree [13]. This lattice, displayed on Fig.1b, contains no closed loops. It can be constructed in the way that starting from q^N $(q+1)$ sites one generates new vertices of in-degree q, called the connectivity of the tree. In subsequent steps $q^{N-1}(q+1)$, $q^{N-2}(q+1),\ldots,$ q+1 vertices are created and finally a vertex of in-degree q+1 is put in the center of the lattice.

(a)

(b)

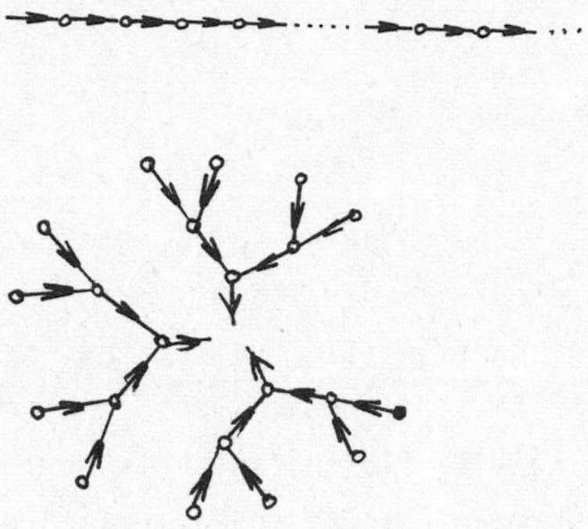

Fig. 1. Iterations along the one dimensional chain (a) and the Cayley tree (b). Here a tree of connectivity q = 2 is depicted. The arrows show the direction of the process.

The iteration is defined by

$$x_{n+1} = \varepsilon\zeta_{n+1} + \sum_{i=1}^{q} f(v_n^{(i)}, x_n^{(i)}) \; . \tag{4.1}$$

Here the subscript n is the index of different generations, being zero for the points on the surface of the tree, where the iteration starts. In Eq.(4.1) a given point of the graph is considered, for which the index i distinguishes different ingoing edges. With each edge we associate the uncorrelated random numbers $v_n^{(i)}$. Provided the process is ergodic, after many iteration steps on a large Cayley tree the probability density for x will get close to the stationary one P(x). The equation for the stationary density P(x) of the iteration (4.1) is given by

$$P(x)=\int\delta(x-\sum_{i=1}^{q} f(v_i,x_i) - \varepsilon\zeta)\rho(\zeta)\prod_{i=1}^{q}(P(x_i)\omega(v_i)dx_i dv_i)d\zeta \quad . \tag{4.2}$$

Starting from this relation, a perturbative expansion can be elaborated for the momenta of x. Instead of treating this case in a general manner, let us turn to a special example.

We consider an Ising model on the Cayley tree, expected to simulate a system of large dimensionality. Note that the total number of lattice sites $M_t = 1+(q+1)(q^{N-1}-1)/(q-1)$ and that of the surface sites $M_s = q^N(q+1)$ are asymptotically equal, i.e. $\lim_{M_t\to\infty} M_t/M_s = 1$. On the other hand, for a lattice of D dimensions $0 < \lim_{M_t\to\infty} M_t/M_s^{\alpha} < \infty$ only if $\alpha = D/(D-1)$. Hence one concludes that, loosely speaking, the Cayley tree is a lattice of infinite dimensionality. Let us turn now to the Hamiltonian of the disordered Ising model

$$-\frac{\mathcal{K}}{k_BT} = \sum_{i,j} K_{ij}\, s_i s_j + \sum_{i,j} h_i s_i \quad , \tag{4.3}$$

where the first sum goes over all neighboring spins, the exchange couplings K_{ij} being uncorrelated and distributed according to a symmetric density $\nu(K)$. The other symbols were defined already in Section III. The physical properties of the model are determined by the iteration (4.1) if one inserts the formulae (3.2) therein. We shall be interested in the spatial fluctuations of the local magnetization far from the surface of the Cayley tree. This is characterized by Q as defined in (3.4). Note that now an average should also be taken over the coupling K. Adopting the technique outlined in Section II for Eq.(4.2), we obtain in the lowest order

$$Q = (1+\langle v^2\rangle)\langle x^2\rangle + \ldots = \frac{1+\langle v^2\rangle}{1-q\langle v^2\rangle}\langle h^2\rangle + \ldots \tag{4.4}$$

where the notation $v = \tanh K$ is used. In contrast to the one-dimensional case, there is in general a finite temperature for which $q\langle v^2\rangle = 1$. To study this critical situation the systematic method of Section II can be used again and it turns out that

$$Q = (q^{-1}+1)\ \langle x^2\rangle + \ldots = (q+1)\left[\frac{\langle h^2\rangle}{2q(q-1)}\right]^{\frac{1}{2}} + \ldots \quad . \tag{4.5}$$

It is noteworthy that the same 'critical exponent' 1/2 arises for $\langle x^2 \rangle$ in Eqs.(2.8) and (4.5). Nevertheless, we did not need here the Kramers-Moyal technique because of the random behaviour of the parameter v. We mention, that for $q\langle v^2 \rangle > 1$ nonzero spatial fluctuations of the local magnetization can be observed even in the absence of magnetic field. Up to first order

$$Q = (q\langle v^2 \rangle - 1)2(q + 1)/(q - 1) + \dots \tag{4.6}$$

is obtained for $1 \gg q\langle v^2 \rangle - 1 > 0$.

V. FULLY DEVELOPED CHAOTIC ONE-DIMENSIONAL MAPS

The main part of this Section is devoted to maps of the interval [0,1] possessing a single maximum, whereas at the end we briefly mention an iteration exhibiting chaos on the entire real axis. First the fully developed chaotic situation [14] is considered (see Fig.2), meaning that the iteration $x_{n+1} = f(x_n)$ is ergodic according to a well-behaved density function P(x) on the whole interval [0,1] and f(0) = = f(1) = 0. Let the function f be C^∞ except possibly at its maximum point. The requirement of ergodicity is met if, for instance, f has a negative Schwarzian derivative [15] or if it is everywhere expanding [16]. Let us restrict our investigations to symmetric f(x) = f(1-x). The explicit form of the probability density function P(x) is only known in some particular cases as e.g. for the hat function

$$f_H(x) = 1 - |1-2x| \quad , \qquad P_H(x) \equiv 1 \ , \tag{5.1}$$

and for the logistic map

$$f_L(x) = 1 - (1 - 2x)^2 \ , \qquad P_L(x) = \frac{1}{\pi\sqrt{x(1-x)}} \quad . \tag{5.2}$$

These maps are smoothly conjugated to each other, which transformation connects all symmetric fully developed chaotic maps each possessing a symmetric probability density function (see [17]). Once a smooth conjugating function to f_H is found, the density of the iteration in question can be given immediately. Nevertheless, the density associated with a symmetric fully developed chaotic map is in general nonsymmetric, hence one cannot determine it by means of conjugation to f_H.

In what follows we shall investigate the invariant measure $\mu(x) = \int_0^x P(y)dy$ of the interval [0,x]. It satisfies an equation which can be read off from Fig. 2

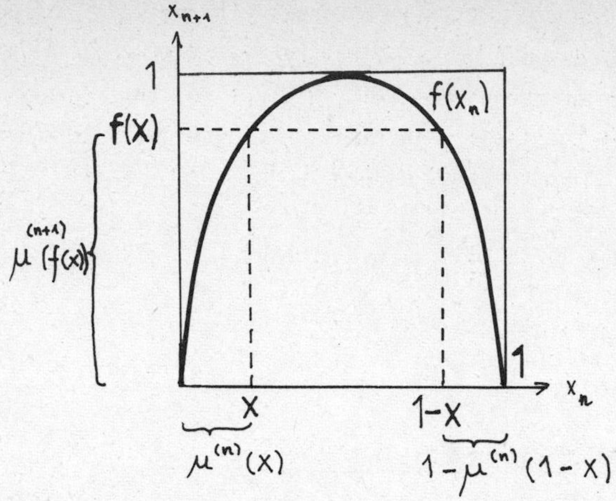

Fig. 2. A symmetric fully developed chaotic map, inducing the transformation of a measure $\mu^{(n)}(x)$ into $\mu^{(n+1)}(x)$.

$$\mu(f(x)) = 1 + \mu(x) - \mu(1 - x) , \qquad x \leq 1/2 \quad . \tag{5.3}$$

The sign of $\mu(x) - \mu(1-x)$ changes if $x > 1/2$. This functional equation cannot generally be solved for an arbitrary symmetric map f. If the map f, however, is not far from a map f_0 with a known, analytic invariant measure $\mu_0(x) = \int_0^x P_0(y)dy$ we can proceed as follows [14]. Consider the map

$$f(\varepsilon;x) = f_0(x) + \varepsilon f_1(x) \tag{5.4}$$

exhibiting fully developed chaos for a finite region of the small parameter ε around $\varepsilon = 0$. We then assume that the corresponding invariant measure can be expanded as

$$\mu(\varepsilon;x) = \sum_{i=0}^{\infty} \mu_i(x)\, \varepsilon^i \quad . \tag{5.5}$$

(The conditions for the analyticity of $\mu(\varepsilon;x)$ in ε are given rigorously in [18].) Using Eqs.(5.3) - (5.5) we obtain functional equations for the functions $\mu_i(x)$. Theoretically, one solves the equation for $\mu_1(x)$.

$$\mu_1(f_0(x)) = -P_0(f_0(x))f_1(x) + \mu_1(x) - \mu_1(1-x) , \quad x < 1/2 , \tag{5.6}$$

then uses the result for writing down the equation for $\mu_2(x)$ which has to be solved next, etc. Unfortunately, in the general case these equations are as hard to solve as Eq.(5.3) is. We are facing a simpler situation, however, if f_0 is the logistic map (5.2) and the symmetric perturbation εf_1 is a polynomial. Then $\mu_0(x) = \frac{2}{\pi}\arcsin\sqrt{x}$ it can be shown that

$$\mu_i(x) = \sqrt{x(1-x)}\, R_i(x) , \tag{5.7}$$

where R_i is a polynomial of x [14]. In other words, the functional equations could be reduced to a comparison of coefficients of polynomials.

As an example we take the quartic map

$$f(\varepsilon;x) = 1 - (1 - \varepsilon)(1 - 2x)^2 - \varepsilon(1 - 2x)^4 , \tag{5.8}$$

for which the density can be calculated up to any finite order:

$$P(\varepsilon;x) = \frac{\partial\mu(\varepsilon;x)}{\partial x} = \frac{1}{\pi\sqrt{x(1-x)}} \{1+\varepsilon\,(x-\tfrac{1}{2}) + \varepsilon^2\,\tfrac{3}{4}(1 - 5x + 4x^2) + \ldots\} . \tag{5.9}$$

The asymmetry of the density is already apparent from the $O(\varepsilon)$ term. The result (5.9) enables us to give explicit formulae for the characteristics of the chaotic process. The Lyapunov exponent and the correlation function read now as

$$\lambda = < \ln \left|\frac{\partial f(\varepsilon;x)}{\partial x}\right| > = \ln 2 - \frac{\varepsilon^2}{16} + \ldots . \tag{5.10}$$

and

$$C(\varepsilon;\tau) = <(x - <x>)(f^{(\tau)}(\varepsilon;x) - <x>)> =$$

$$= \frac{1}{8}\delta_{\tau,0} - \varepsilon 2^{-5}\delta_{\tau,1} + \varepsilon^2(2^{-7}\delta_{\tau,0} + 3.2^{-7}\delta_{\tau,1} + 2^{-6}\delta_{\tau,2}) + \ldots , \tag{5.11}$$

respectively. Here <> means the ergodic average, $f^{(\tau)}(\varepsilon;x)$ is the τ-th iterate of x, and $\delta_{\alpha,\beta} = 1$ if $\alpha = \beta$, otherwise it is zero. Apparently for $\varepsilon = 0$ we get back a delta-correlated process with the maximal Lyapunov exponent $\lambda = \ln 2$. The latter properties, characterizing all symmetric maps with symmetric densities, indicate a high degree of erraticity. Eqs. (5.10) and (5.11) just display the way the degree of randomness decreases. By using the above results, the effect of a small, time-dependent perturbation on the iteration of the quartic map can be calculated explicitly [19,20]. We mention that a systematic technique has be developed around $f_H(x)$ by the authors of [14] and independently in [18]. In particular, for

$$\tilde{f}(\varepsilon;x) = f_H(x) + \varepsilon f_H(x)(1 - f_H(x)) \tag{5.12}$$

the probability density function is given by

$$\tilde{P}(\varepsilon;x) = 1 + \varepsilon(2x - 1) + \varepsilon^2(7/3 - 10x + 8x^2) + \ldots \quad . \tag{5.13}$$

Finally we briefly discuss recent results [21] on a chaotic iteration

$$x_{n+1} = (E - \varepsilon\zeta_n - x_n)^{-1} \quad , \qquad |E| \leq 2 \quad , \tag{5.14}$$

covering the entire real axis. This process is a transformed version of the Schrödinger equation for a disordered quantum mechanical system on a chain (see e.g.[22]). Here E is the energy of the real valued eigenvector ψ_n, $x_n = \psi_{n-1}/\psi_n$, and ζ_n are uncorrelated random numbers. It turns out that for the particular energies $E = 2\cos\pi\alpha$, α rational, a systematic expansion for the probability density $P(E,\varepsilon;x)$ according to ε can be carried out, and its energy-derivatives can also be given [21]. Among the ergodic averages, which also can be given in form of series, the localization length $\zeta(E,\varepsilon) = -1/\langle \ln x^2 \rangle$ is of special interest, characterizing on the average the exponential decay of the state ψ_n of energy E for large distances (see e.g.[23]). In order to gain some insight how the perturbative method works let us consider the case E = 0. Then supposing a form

$$P(E = 0,\varepsilon;x) = \sum_{i=0}^{\infty} P_i(x)\varepsilon^i \tag{5.15}$$

one obtains from the invariance condition for $P(0,\varepsilon;x)$

$$x^2\, P_0(x) = P_0(-1/x) \ , \tag{5.16a}$$

$$x^2\, P_1(x) = P_1(-1/x) \ , \tag{5.16b}$$

$$x^2\, P_2(x) = P_2(-1/x) + P_0''(-1/x)/2 \ , \tag{5.16c}$$

and so on. Since Eq.(5.16a) does not determine P_0 uniquely, one should also use Eq.(5.16c). After some calculations one arrives at

$$(1+x^4)\, P_0''(x) + 6x^3\, P_0'(x) + 6x^2\, P_0(x) = 0 \;, \tag{5.17}$$

which yields the normalized solution

$$P_0(x) = 2\Gamma(\tfrac{1}{2})\{\Gamma(\tfrac{1}{4})\}^{-2}\, (1 + x^4)^{-1/2} \quad . \tag{5.18}$$

This enables us to calculate the localization length as

$$\xi^{-1} = -\varepsilon^2\langle\zeta^2\rangle \int P_2(x) \ln x^2\, dx + \ldots = 2\varepsilon^2\langle\zeta^2\rangle\{\Gamma(\tfrac{3}{4})/\Gamma(\tfrac{1}{4})\}^2 + \ldots \;, \tag{5.19}$$

where Eqs.(5.16c), (5.18) have been used [21]. It is noteworthy that while the process (5.14) is not chaotic for $\varepsilon = 0$, yet the density extends over the entire real axis in the limit $\varepsilon \to 0$. In order to calculate higher order terms P_i one has to take into account equations for E- and ε-derivatives of $P(E,\varepsilon;x)$ as well.

Summarizing the above findings, two examples have been shown where the functional equation for a density could be transformed into an infinite sequence of local equations, solvable step by step. The transformations, however, turned out to be very different.

REFERENCES

[1] Haken, H., (ed.): Chaos and Order in Nature, Springer(1981).

[2] Haken, H., (ed.): Evolution of Order and Chaos, Springer(1982)

[3] Garrido, L. (ed.): Dynamical Systems and Chaos, Lecture Notes in Physics 179 Springer (1983).

[4] Eilenberger, G., Müller-Krumbhaar, H. (ed.): Nichtlineare Dynamik in kondensierter Materie, KFA Institut für Festkörperforschung, Jülich (1983).

[5] Cohen, E.G.D. (ed.): Proceedings of the NUFFIC Summer School on Statistical Mechanics, Trondheim (1984), to appear.

[6] Kramers, H.A.: Physica 7, 284 (1940).

[7] Moyal, J.E.: J.R. Stat. Soc. 11, 151 (1949).

[8] Bruinsma, R., Aeppli, G.: Phys. Rev. Lett. 50, 1494 (1983).

[9] Bruinsma, R., Aeppli, G.: Phys. Lett. 97A, 177 (1983).

[10] Györgyi, G., Rujân, P.: J. Phys. C: Solid St. Phys. 17, 4207 (1984).

[11] Sütő, A., Zimânyi, G.T.: Lecture Notes in Physics 206, Springer (1984).

[12] Derrida, B., Pomeau Y., Vannimenus, J.: J. Phys. C: Solid St. Phys. 11, 4749 (1978).

[13] Cayley, A.: Coll. Math. Papers 3, 242 (1889).

[14] Györgyi, G., Szépfalusy, P.: J. Stat. Phys. 34, 451 (1984).

[15] Misiurewicz, M.: Publ. Math. IHES 53, 17 (1981).

[16] Lasota, A., Yorke, J.A.: Trans. Am. Math. Soc. 183, 481 (1973).

[17] Grossmann, S., Thomae, S.: Z. Naturforsch. 32a, 1353 (1977).

[18] Keller, G.: Analytic Perturbations of N-Fold Covering Interval Mappings, (1984) unpublished.

[19] Geisel, T., Heldstab, J., Thomas, H.: Z. Phys. B55, 165 (1984).

[20] Grossmann, S.: Z. Phys. B57, (1984) to appear.

[21] Derrida, B., Gardner, E.: J. de Phys. 45, 1283 (1984).

[22] Ishii, K.: Sup. Prog. Theor. Phys. 53, 77 (1973).

[23] Sarker, S.: Phys. Rev. B25, 4304 (1982).

Géza Györgyi
Institute for Theoretical Physics
Eötvös University
H-1088 Budapest, P. O. B. 327

ITINERARIES UNDER UNIMODAL MAPS

Gilbert Helmberg

Let f be a unimodal map of [-1,+1] into [-1,+1] in the sense of [1], i.e. f is continuous, strictly increasing on [-1,0], strictly decreasing on [0,+1], and f(0) = 1. We shall refer to the "piecewise linear" case if f is linear on [-1,0] and on [0,+1] with slopes larger than 1.

We shall adhere to the terminology used in [1]. In particular we define I(x) to denote the letter L,C or R according to whether x is smaller than, equal to, or larger than 0. The "itinerary" $\underline{I}(x)$ is the finite sequence $\{I(f^{k-1}(x)\}_{k=1}^{n}$ if $f^{n-1}(x) = 0$ and if $n \in \mathbb{N}$ is minimal, and the infinite sequence $\{I(f^{k-1}(x))\}_{k=1}^{\infty}$ if $f^{k-1}(x) \neq 0$ for all $k \in \mathbb{N}$. Correspondingly we call a word $\underline{A}$ from the alphabet {L,C,R} "admissible" if it either stops with the first appearance of the letter C or if it is infinite but does not contain the letter C at all. By the "length" $|\underline{A}|$ of $\underline{A}$ we denote the number of letters in $\underline{A}$ (which may be infinite).

In this note we are concerned with the question which admissible words $\underline{A}$ "appear" as an itinerary under a given unimodal map f (i.e. there exists an $x \in [-1,+1]$ such that $\underline{I}(x) = \underline{A}$). The intention is to complete (at least in some reasonable sense) the partial answer to this question given in [1] theorem II.3.8.

Recall that finite words are either "even" or "odd" in correspondence with the parity of the number of letters R which they contain. Moreover, one has an order relation < defined in the set $\mathcal{A}$ of admissible words, defined by L < C < R and $\underline{A} = X_1 \dots X_n A_{n+1} \dots < \underline{B} = X_1 \dots X_n B_{n+1} \dots (A_{n+1} \neq B_{n+1})$ iff $\underline{X} = X_1 \dots X_n$ is even (resp. odd) and $A_{n+1} < B_{n+1}$ (resp. $A_{n+1} > B_{n+1}$). This order reflects the order of points in [-1,+1] inasfar as x < y implies $\underline{I}(x) \leq \underline{I}(y)$. A non-degenerate subinterval of [-1,+1] is called a homterval if $\underline{I}(x)$ is constant for all of its points. Homtervals cannot exist e.g. in the piecewise linear case but do exist if f is C^1-unimodal (i.e. f is C^1 besides unimodal) and $\underline{I}(1)$ is finite ([1] lemma II.3.9).

Let the "left shift operator" S be defined on $\mathcal{A}$ by $S\underline{A} = S(A_1A_2\dots) = A_2A_3\dots$. For all $x \in [-1,+1]$ one has $\underline{I}(f(x)) = S\underline{I}(x)$ and, putting $\underline{B} = \underline{I}(x)$,

$$\underline{I}(-1) \leq S^k\underline{B} \leq \underline{I}(1) \quad \text{for all} \quad k < |\underline{B}| \ . \tag{1}$$

In order to simplify the statement of [1] theorem II.3.8, given a finite word $\underline{A} = A_1 \dots A_{n-1}A_n$ we denote by $\underline{A}^+, \underline{A}^-$, and $\underline{A}^0$ respectively the even, odd (non-admissible), and admissible one of the three words $A_1 \dots A_{n-1}X$ ($X \in \{L,R,C\}$). Then a word $\underline{B} \in \mathcal{A}$ appears as an itinerary (under f) if

$$\underline{I}(-1) \leq S^k\underline{B} < \underline{I}(1) \quad \text{for all} \quad k < |\underline{B}| \quad \text{if} \quad \underline{I}(1) \quad \text{is infinite,}$$
$$\underline{I}(-1) \leq S^k\underline{B} < (\underline{A}^+)^\infty \quad \text{for all} \quad k < |\underline{B}| \quad \text{if} \quad \underline{I}(1) = \underline{A}^0 \quad \text{is finite.} \tag{2}$$

The object of the present note is a clarification of the space left between (1) and (2): what can be said about appearance of words $\underline{B} < \underline{I}(1)$ satisfying (1) but for some positive $k < |\underline{B}|$ also

$$S^k\underline{B} = \underline{I}(1) \qquad \text{if } \underline{I}(1) \quad \text{is infinite or} \tag{3}$$

$$(\underline{A}^+)^\infty \leq S^kB \leq \underline{I}(1) \qquad \text{if } \underline{I}(1) = A^0 \quad \text{is finite.} \tag{4}$$

To this end recall that the "kneading sequence" $\underline{M} = \underline{I}(1)$ is "maximal" in the sense that $S^k\underline{M} \leq \underline{M}$ for all $k < |\underline{M}|$. Similarly $\underline{H} = \underline{I}(-1)$ is "minimal", i.e. satisfies $S^k\underline{H} \geq \underline{H}$ for all $k < |\underline{H}|$. Given any maximal word $\underline{M} \in A$ we define the set of admissible words "weakly dominated by $\underline{M}$" by

$$\mathcal{D}^0(M) = \{\underline{B} \in A : S^k\underline{B} \leq \underline{M} \text{ for all } k < |\underline{B}|\} .$$

In [2] it is shown that $\mathcal{D}^0(M)$ is compact in the order topology in A and that its greatest cluster point $\underline{\bar{M}} = \limsup \mathcal{D}^0(\underline{M})$ depends on the structure of $\underline{M}$ in the following way. We have to distinguish the four cases:

a) $\underline{M}$ is infinite aperiodic;

b) $\underline{M} = (\underline{A}^+)^\infty$ (i.e. $\underline{M}$ has an even primitive period $\underline{A}^+$);

c) $\underline{M} = \underline{A}^0$ (i.e. $\underline{M}$ is finite);

d) $\underline{M} = (\underline{A}^-)^\infty$ (i.e. $\underline{M}$ has an odd primitive period $\underline{A}^-$).

Then $\underline{\bar{M}} = \underline{M}$ in cases a) and b), and $\underline{\bar{M}} = (\underline{A}^+)^\infty$ in cases c) and d).

For a minimal word $\underline{H} \in A$ let

$$\mathcal{C}^0(\underline{H}) = \{\underline{B} \in A \ : \underline{H} \leq S^k\underline{B} \quad \text{for all} \quad k < |\underline{B}|\} .$$

A routine check reveals that for any (non-empty) minimal word $\underline{H} \in \mathcal{D}^0(\underline{M})$ satisfying $\underline{H} \leq S\underline{M}$ one has

$$S\{\underline{B} \in \mathcal{D}^0(\underline{M}) : \underline{B} \geq \underline{H}\} \subset \{\underline{B} \in \mathcal{D}^0(\underline{M}) : \underline{B} \geq \underline{H}\}$$

and

$$\underline{H} \leq \underline{K} = \inf \{S^k\underline{M} : k < |\underline{M}|\} \leq \underline{\bar{M}} .$$

In particular this implies

$$C^0(\underline{H}) \cap \mathcal{D}^0(\underline{M}) = \{\underline{B} \in \mathcal{D}^0(\underline{M}) : \underline{B} \geq \underline{H}\} \quad . \tag{5}$$

Also, if $\underline{H} = S\underline{B}$ for some word $\underline{B} \in \mathcal{D}^0(\underline{M})$, then either $\underline{B} = \underline{H} = L^\infty$ (which is the absolutely smallest word) or $\underline{B} = \underline{M}$. By (1) all itinearies belong to $C^0(\underline{H}) \cap \mathcal{D}^0(\underline{M})$ for $\underline{H} = \underline{I}(-1)$ and $\underline{M} = \underline{I}(1)$.

Note that the words $\underline{B}$ of interest for us by (3) and (4) satisfy $S^k\underline{B} \geq \underline{\bar{M}}$ for some $k \in \mathbb{N}$. We write

$$\mathcal{L}^o_{\underline{M}} = \{\underline{B} \in \mathcal{D}^0(\underline{M}) : \underline{\bar{M}} \leq \underline{B} \leq \underline{M}\} \quad .$$

<u>Theorem:</u> Let f, $\underline{M} = \underline{I}(1)$ and $\underline{H} = \underline{I}(-1)$ be given.

A) Suppose $\underline{B} = \underline{I}(x) < \underline{M}$ satisfies (3) (in cases a) b))d) or (4) (in case c)). Let $\underline{B} = \underline{B}_{(k)}S^k\underline{B}$ where $k \geq 0$ is chosen to be minimal such that $S^k\underline{B} \geq \underline{\bar{M}}$. Then either x belongs to a homterval or $\underline{B} = \underline{H}, \underline{B}_{(k)}$ is odd, and $f^k(x)$ belongs to a homterval. Furthermore

1) $S^k\underline{B} = \underline{\bar{M}}$ in cases a), b), c) and $S^k\underline{B} = (\underline{A}^+)^n(\underline{A}^-)^\infty$ for some $n \geq 0$ in case d);
2) if $k > 0$ then $\underline{B}^o_{(k)} \in \mathcal{D}^0(\underline{\bar{M}})$; moreover, if $\underline{B} \neq \underline{H}$, then $\underline{B}^o_{(k)}$ (in cases a) b) c) and d), n = 0) respectively $\underline{B}_{(k)}(\underline{A}^+)^{n-1}\underline{A}^0$ (in case d), n > 0) appears as an itinerary.

B) If $\underline{B}^o_{(k)} \in C^0(\underline{H}) \cap \mathcal{D}^0(\underline{\bar{M}})$ is given, then all words in the set

$$\mathcal{L}^o_{\underline{M}}(\underline{B}^o_{(k)}) = \underline{B}^+_{(k)}\mathcal{L}^o_{\underline{M}} \cup \{\underline{B}^o_{(k)}\} \cup \underline{B}^-_{(k)}\mathcal{L}^o_{\underline{M}} \quad \text{(in cases b) d))}$$

$$\{\underline{B}^+_{(k)}(\underline{A}^+)^\infty, \underline{B}^o_{(k)}, \underline{B}^-_{(k)}(\underline{A}^+)^\infty\} \quad \text{(in case c) if f is } C^1\text{-unimodal)}$$

appear as itineraries unless $\underline{H}$ belongs to this set, in which case precisely the words in this set not smaller than $\underline{H}$ appear as itineraries.

<u>Proof.</u> First we recall some relevant facts from [2]. In the non-trivial cases c) d) (where $\underline{\bar{M}} < \underline{M}$) the chain $K^o_{\underline{M}}$ of inequalities between consecutive words of $\mathcal{L}^o_{\underline{M}}$ has the form

$$\underline{\bar{M}} = (\underline{A}^+)^\infty < \ldots < (\underline{A}^+)^n\underline{A}^0 < (\underline{A}^+)^{n-1}\underline{A}^0 < \ldots < \underline{A}^0 = \underline{M} \qquad \text{(case c))} \tag{6}$$

$$\underline{\bar{M}} = (\underline{A}^+)^\infty < \ldots < (\underline{A}^+)^n\underline{A}^0 < (\underline{A}^+)^n(\underline{A}^-)^\infty < (\underline{A}^+)^{n-1}\underline{A}^0 < \ldots$$
$$\ldots < \underline{A}^+(\underline{A}^-)^\infty < \underline{A}^0 < (\underline{A}^-)^\infty = \underline{M} \qquad \text{(case d))} \tag{7}$$

([2] theorem 8.5). If $\underline{B} \in \mathcal{D}^0(\underline{M})$ satisfies $S^k\underline{B} \geq \underline{\bar{M}}$ (as in (3) and (4)) for some minimal $k \in \mathbb{N}$, then writing $\underline{B} = \underline{B}_{(k)}S^k\underline{B}$ as in the theorem one has $\underline{B}^0_{(k)} \in \mathcal{D}^0(\underline{\bar{M}})$ ([2] lemma 8.6) and $\underline{B} \in L^0_{\underline{M}}(\underline{B}^0_{(k)}) \subset \mathcal{D}^0(\underline{M})$ in all cases a) b) c) d). Also, there are no words in $\mathcal{D}^0(\underline{M})$ between consecutive ones in the chain of inequalities

$$\underline{B}^+_{(k)}K^0_{\underline{M}} < \underline{B}^0_{(k)} < \underline{B}^-_{(k)}K^0_{\underline{M}} \tag{8}$$

(preposition of the odd word $\underline{B}^-_{(k)}$ reverses the inequalities in $K^0_{\underline{M}}$, and $K^0_{\underline{M}}$ "degenerates" to $\underline{\bar{M}} = \underline{M}$ in cases a) b) [2] theorem 8.7). We shall refer to this fact as the "gap property" of the chain (8). Conversely, for any word $\underline{B}^0_{(k)} \in \mathcal{D}^0(\underline{\bar{M}})$ one has $\underline{B}^{\pm}_{(k)}L^0_{\underline{M}} \subset \mathcal{D}^0(\underline{M})$ ([2] lemma 8.6).

Consider now $\underline{B} = \underline{B}_{(k)}\, S^k\underline{B} = \underline{I}(x)$ as in assertion A). By our assumption $\underline{B} < \underline{M}$, k can be zero only in cases c) and d). A finite word can only appear as the itinerary of one single point. By (6), the only words in $L^0_{\underline{M}}$ which can appear at all in case c) are $\underline{A}^0$ and $(\underline{A}^+)^\infty$, the last one necessarily only on a homterval. Also, for $k > 0$ in case c), $S^k\underline{B} = \underline{A}^0$ is excluded since $\underline{A}^0$ is the itinerary of 1 only. In case d) the word $(\underline{A}^+)^\infty$ and every finite word in $L^0_{\underline{M}}$ appears as an itinerary by (2) and (7), consequently so do the remaining intermediate words of the form $(\underline{A}^+)^n(\underline{A}^-)^\infty$, again necessarily on homtervals. This also demonstrates point 1 of assertion A).

The next lemma serves to cope with a non-zero k.

Lemma. Let $\underline{B} = \underline{B}_{(k)}S^k\underline{B}$ as in assertion A). If $k > 0$ and $S^l\underline{B}^0_{(k)} < \underline{H}$ for some $l < k$ then $l = 0$. Furthermore, in cases a) b) c) one then has $\underline{B} = \underline{H} = \underline{B}_{(k)}\underline{\bar{M}}$, and $[f^k(-1),1]$ is a homterval.

Proof. The inequality

$$S^l\underline{B}^0_{(k)} < \underline{H} \leq S^l\underline{B} = S^l\underline{B}_{(k)}S^k\underline{B}$$

by the gap property of the chain (8) implies that $S^l\underline{B}_{(k)}$ has to be odd and

$$\underline{H} = S^l\underline{B} = S^l\underline{B}_{(k)}\underline{\bar{M}} \text{ in cases a) b) c)}$$

$$\underline{H} = S^l\underline{B}_{(k)}\underline{D} \quad \text{for some } \underline{D} \text{ in } L^0_{\underline{M}} \text{ in case d)} \quad .$$

If $l > 0$ then $\underline{H} = S\underline{M} = S^l\underline{B}_{(k)}S^k\underline{B}$ (by the remark preceding the theorem) implies

$$\underline{M} \in S^{l-1}\underline{B}_{(k)}L^0_{\underline{M}} \quad .$$

This is apriori impossible in cases a) and c). In cases b) and d) it would contradict the minimality of k. Consequently one has $l = 0$. In cases a) b) c) one obtains

$\underline{I}(f^k(-1)) = S^k\underline{H} = \bar{\underline{M}}$. Therefore $\underline{H}$ is infinite and one has $f^{k-1}(-1) \neq 0$ and $f^k(-1) \neq$ $\neq 1$. In cases a) b) both $f^k(-1)$ and 1 have the same itinerary. In case c) the gap property of the chain (6) implies that $[f^k(-1),1[$ is a homterval. This proves the lemma.

Returning to assertion A) in cases a) b) c), $k > 0$, by the lemma we have either $\underline{B} = \underline{H}$, $\underline{B}_{(k)}$ odd, and $f^k(x)$ in the homterval with itinerary $S^k\underline{H} = \bar{\underline{M}}$, or $S^l\underline{B}^o_{(k)} \geq \underline{H}$ for every $l < k$. In this last case $\underline{B}^o_{(k)} \in C^o(\underline{H}) \cap \mathcal{D}^o(\bar{\underline{M}})$ appears as an itinerary by (2) and $\underline{B} = \underline{I}(x)$ defines a homterval by the gap property of the chain (8).

In case d) the "inner neighbour" $(S^l\underline{B})^\sim$ of $S^l\underline{B} = S^l\underline{B}_{(k)}(\underline{A}^+)^n(\underline{A}^-)^\infty$ in the chain (8) is $S^l\underline{B}^o_{(k)}$ for $n = 0$ and $S^l\underline{B}_{(k)}(\underline{A}^+)^{n-1}\underline{A}^o$ for $n > 0$. This shows $(S^l\underline{B})^\sim = S^l(\tilde{\underline{B}})$. The inequality $S^l\tilde{\underline{B}} < \underline{H}$ is possible only if

$$S^l\underline{B}^o_{(k)} \leq S^l\tilde{\underline{B}} < \underline{H} = S^l\underline{B} .$$

This implies $l = 0$ by the lemma, and $\underline{B}_{(k)}$ has to be odd. Moreover $\underline{I}(f^k(x)) =$ $= (\underline{A}^+)^n(\underline{A}^-)^\infty$ defines a homterval by the gap property of the chain (7). If $\underline{H} \neq \underline{B}$ then $\tilde{\underline{B}} \in C^o(\underline{H}) \cap \mathcal{D}^o(\underline{M})$ appears as an itinerary by (2). By the gap property of the chain (8) the word $\underline{B} = \underline{I}(x)$ defines a homterval again.

If the kneading sequence $\underline{M}$ is periodic (as in cases b) d)) with period length m, then $f^{m-1}(1) \neq 0$ and $f^m(1) \neq 1$ but $\underline{I}(f^m(1)) = S^m\underline{M} = \underline{M}$; therefore $[f^m(1),1]$ is a homterval. Similarly in case c) if f is C^1-unimodal, then $\bar{\underline{M}} = (\underline{A}^+)^\infty$ defines a homterval. Under the hypothesis of assertion B), the word $\underline{B}^o_{(k)}$ has to appear as an itinerary by (2). Then for adjoining homtervals there also have to appear the itineraries $\underline{B}^\pm_{(k)}\underline{M}$ (in cases b) d)) resp. $\underline{B}^\pm_{(k)}\bar{\underline{M}}$ (in case c)) unless $\underline{H} = \underline{B}^o_{(k)}$ in which case only $\underline{B}^-_{(k)}\underline{M}$ resp. $\underline{B}^-_{(k)}\bar{\underline{M}}$ appears. The equation $(S^l\underline{B}_{(k)})^+\underline{D} = \underline{H} = S\underline{M}$ for some positive $l < k$ and some $\underline{D} \in L^o_{\underline{M}}$ is impossible in case c) by (6). In cases b) d) $S^{l-1}\underline{B}^o_{(k)} \neq \underline{A}^o$ implies $S^{l-1}\underline{B}^o_{(k)} = LS^l\ \underline{B}^o_{(k)}$. By the gap property of the chain (8) one obtains

$$S^{l-1}\underline{B}^o_{(k)} = LS^l\underline{B}^o_{(k)} > S^l\underline{B}^o_{(k)} > \underline{H} = S\underline{M}$$

and "cancelling" the initial letter L

$$S^l\underline{B}^o_{(k)} = S\underline{A}^o > S^{l+1}\underline{B}^o_{(k)} = S^2\underline{A}^o > S\underline{H} \geq \underline{H} = S\underline{M} \quad .$$

This is again impossible by the same gap property. Therefore $l = 0$ and consequently also in case d) the word $\underline{B}^+_{(k)}(\underline{A}^+)^n(\underline{A}^-)^\infty$ appears as long as $\underline{H} \leq B^+_{(k)}(\underline{A}^+)^n(\underline{A}^-)^\infty$. This completes the proof of the theorem.

The following observations may be seen as corollaries:

1) A unimodal map with periodic kneading sequence has homtervals.
2) If homtervals are excluded (as in the piecewise linear case or in case of an S-unimodal map with aperiodic kneading sequence by [2] theorem II.5.4 and proposition II.6.2), then a word $\underline{B} < \underline{I}(1)$ appears as an itinerary iff it satisfies (2)
3) The infinite words $\underline{B} \in C^{o}(H) \cap \mathcal{D}^{o}(\underline{M})$ which do not appear as itineraries are (order topology) limits of finite words in $C^{o}(\underline{H}) \cap \mathcal{D}^{o}(\underline{M})$ which do appear as itineraries. Therefore these words turn up as limits of itineraries.

REFERENCES

[1] Collet, P., Eckmann, J.P.: Iterated maps on the interval as dynamical systems. Birkhäuser, Boston 1980.

[2] Sun, L., Helmberg, G.: Maximal words connected with unimodal maps. Institutsnotiz Nr.2, Studienjahr 1983/84, Institut für Mathematik und Geometrie, Universität Innsbruck, submitted for publication.

Gilbert Helmberg
Technische Fakultät
Institut für Mathematik
und Geometrie
Technikerstraße 16
A-6020 Innsbruck

CAUCHY FUNCTIONAL EQUATION ON A RESTRICTED DOMAIN AND COMMUTING FUNCTIONS

Janusz Matkowski

INTRODUCTION

We consider the Cauchy functional equation

$$\phi(x+y) = \phi(x) + \phi(y), \qquad (x,y) \in X \subset R^2, \tag{1}$$

where $\phi:R \to R$ is continuous at least at one point and X is either a union of two perpendicular straight lines $x = a$, $y = b$, or a union of two parallel straight lines $x = a$, $x = b$ (or $y = a$, $y = b$), such that a and b are not commensurable. It is easily seen that equation (1) can be written as the following system of two simultaneous functional equations

$$\phi(t+a) = \phi(t) + \phi(a), \quad \phi(t+b) = \phi(t) + \phi(b), \quad t \in R. \tag{2}$$

Theorem 1 says that there is a $c \in R$ such that $\phi(t) = ct$, $t \in R$. This result enables us to prove Theorem 2 which reads as follows. Suppose that $(f^t)_{t \in R}$ is a measurable iteration group of a function $f:I \to I$ which is strictly increasing, onto and without fixed points in the interior of an interval I. If $g:I \to I$ is continuous at least at one point of int(I) and g commutes with two functions f^a and f^b then $g = f^c$ for some $c \in R$.

Using this Theorem we give a contribution to the result of Jens Schwaiger concerning iteration semigroups which are commuting in pairs.

1. WE BEGIN WITH THE FOLLOWING

Theorem 1. Suppose that $\phi:R \to R$ satisfies (2) where $a,b \in R$, $a \neq 0$, and b/a is irrational. If ϕ is continuous at least at one point then there is a $c \in R$ such that $\phi(t) = ct$ for $t \in R$.

Proof. Suppose first that ϕ is continuous at $t = 0$. It follows easily from (2) that

$$\phi(t+na) = \phi(t)+n\phi(a), \ \phi(t+mb) = \phi(t)+m\phi(b), \quad n,m \in Z, \ t \in R, \tag{3}$$

which implies that

$$\phi(t+na+mb) = \phi(t)+n\phi(a)+m\phi(b), \quad n,m \in Z, \quad t \in R. \tag{4}$$

Because $\phi(0) = 0$ we hence get

$$\phi(na+mb) = n\phi(a)+m\phi(b), \quad n,m \in Z. \tag{5}$$

The set

$$A = \{na+mb: n,m \in Z\}$$

is dense in R (cf.Halmos [1], p.69, Theorem G). Therefore there exists a sequence $t_k = n_k a+m_k b \in A$, $k = 1,2,\ldots$, such that

$$\lim_{k\to\infty} t_k = 0, \quad t_k = n_k a+m_k b \neq 0, \quad k \in N. \tag{6}$$

From (5) by the continuity of ϕ at the point $t = 0$ we have

$$\lim_{k\to\infty} (n_k\phi(a)+m_k\phi(b)) = \phi(0) = 0 . \tag{7}$$

It follows from (6) that $m_k \neq 0$ for all $k \in N$ and

$$\lim_{k\to\infty} \frac{n_k}{m_k} = -\frac{b}{a} .$$

Now (7) implies that

$$\lim_{k\to\infty} \frac{n_k\phi(a)+m_k\phi(b)}{m_k} = -\frac{b}{a}\phi(a)+\phi(b) = 0 ,$$

i.e., $\frac{\phi(a)}{a} = \frac{\phi(b)}{b}$. Consequently, there exists a $c \in R$ such that

$$\phi(a) = ca, \quad \phi(b) = cb.$$

Hence and from (5) we obtain $\phi(na+mb) = c(na+mb)$ for $n,m \in Z$, i.e.

$$\phi(t) = ct, \quad t \in A.$$

Take $t_0 \in R$. Since the set t_0+A is dense in R there exists a sequence $t_k \in A$, $t_k = n_k a+m_k B$, $(k \in N)$, such that

$$\lim_{k\to\infty} (t_0+t_k) = 0, \quad t_k \neq 0, \quad k \in N.$$

Now, according to (4), we have

$$\phi(t_o+n_ka+m_kb) = \phi(t_o)+c(n_ka+m_kb) = c(t_o+n_ka+m_kb)+(\phi(t_o)-ct_o)$$

and the continuity of ϕ at the point $t = 0$ implies that $\phi(t_o) = ct_o$. This proves that $\phi(t) = ct$ for all $t \in R$.

To finish the proof note that if ϕ is continuous at a point $t_o = 0$ and satisfies equations (2) then the function $\psi:R \to R$ defined by the formula $\psi(t) = \phi(t+t_o) - \phi(t_o)$ is continuous at the point $t = 0$ and also satisfies equations (2).

Remark. The same reasoning as in the above proof and observation that $m_k \to \infty$ as $k \to \infty$ allows us to prove that if ϕ is bounded in a neighbourhood of $t = 0$ then $\phi(t) = ct$ for all t belonging to the dense set A. But the boundedness of ϕ in a neighbourhood of $t = 0$ does not imply that $\phi(t) = ct$ for $t \in R$. In fact, $\phi(t) = 0$ for $t \in A$ and $\phi(t) = 1$ for $t \in R-A$ satisfies eq.(2), is bounded and, evidently, it is not linear.

As a simple consequence of Theorem 1 we get the following useful.

Corollary 1. Let $a,b,\alpha,\beta \in R$ and suppose that $\psi:R \to R$ is continuous at least at one point and satisfies the following system of equations

$$\psi(t+a) = \psi(t)+\alpha, \quad \psi(t+b) = \psi(t) + \beta . \tag{8}$$

(a) If $a \neq 0$ and b/a is irrational then there exist $c_o,c_1 \in R$ such that $\psi(t) = c_o + c_1t$ for $t \in R$.

(b) If $\alpha \neq 0$ and β/α is irrational then $a \neq 0$ and b/a is irrational.

Proof. (a) Put $c_o = \psi(0)$ and note that $\phi(t) = \psi(t) - \psi(0)$ satisfies equations (2).

(b) From (8) we have $\psi(na+mb) = \psi(0)+n\alpha+m\beta$, $n,m \in Z$. Now $\alpha \neq 0$ implies $a \neq 0$. If b/a were rational we would find integers n and m, at least one different from 0, such that $na+mb = 0$. Consequently, we would have $\psi(0) = \psi(na+mb) = \psi(0) + n\alpha + m\beta$.

Hence we conclude that $n\alpha+m\beta = 0$ which is a contradiction. For the multiplicative version of the Cauchy equation we get in the same way

Corollary 2. Let a,b,α,β be positive real and suppose that $\psi:(0,\infty) \to (0,\infty)$ is continuous at least at one point and satisfies the following system of simultaneous Schröder functional equations

$$\psi(at) = \alpha\psi(t), \quad \psi(bt) = \beta\psi(t).$$

(a) If $a \neq 1$ and $\ln b/\ln a$ is irrational then there are $c_o > 0$ and $c_1 \in \mathbb{R}$ such that $\psi(t) = c_o t^{c_1}$.

(b) If $\alpha \neq 1$ and $\ln\beta/\ln\alpha$ is irrational then $a \neq 1$ and $\ln b/\ln a$ is irrational.

Remark. Corollary 2 is a generalization of a similar result in [2] where ψ is assumed to be monotonic. This result can be applied to a characterization of L^p norm.

2. IN THIS SECTION WE CONSIDER ITERATION GROUPS AND COMMUTING FUNCTIONS

Let I be an interval. A family of functions $(f^t)_{t\in R}$ is said to be an iteration group of a function $f:I \to I$ iff $f^t:I \to I$, $f^1 = f$, $f^t \circ f^s = f^{t+s}$ for all $t,s \in R$. Replacing in this definition R by the interval $(0,\infty)$ we obtain the definition of an iteration semigroup.

An iteration group $(f^t)_{t\in R}$ (semigroup $(f^t)_{t>0}$) is said to be continous resp. measurable if for every $x \in I$ the mapping $R \ni t \to f^t(x)((0,\infty) \ni t \to f^t(x))$ is continous resp. measurable.

Now we are going to prove the following

Theorem 2. Let $f:I\to I$ be strictly increasing, onto and without fixed points in an open interval I. Suppose that $(f^t)_{t\in R}$ is a measurable iteration group of f with each f^t continuous. If a function $g:I \to I$ continuous at least at one point commutes with two functions f^a and f^b such that b/a is irrational then there exists a $c \in R$ such that $g = f^c$ i.e., g belongs to the iteration group of f.

Proof. According to Zdun's theorem, (f^t) has to be a continous iteration group (cf.[3]). Therefore there exists a function $\alpha:R \to I$ continous, strictly monotonic and onto such that

$$f^t(x) = \alpha(t+\alpha^{-1}(x)), \quad x \in I, \quad t \in R. \tag{9}$$

Using the assumption $f^a \circ g = g \circ f^a$ we have

$$\alpha(a+\alpha^{-1}(g(x))) = g(\alpha(a+\alpha^{-1}(x))), \quad x \in I.$$

Since α is onto, for each $t \in R$ there is an $x \in I$ such that $x = \alpha(t)$. Substituting this into the above equation we get

$$\alpha(a+\alpha^{-1}(g(\alpha(t))) = g(\alpha(a+t)), \quad t \in R.$$

Hence, putting

$$\phi = \alpha^{-1} \circ g \circ \alpha \tag{10}$$

we get the equation

$$\phi(t+a) = \phi(t)+a . \tag{11}$$

Similarly, from $f^b \circ g = g \circ f^b$ we have

$$\phi(t+b) = \phi(t)+b . \tag{12}$$

Since ϕ is continuous at least at one point, in view of Corollary 1 with $\alpha = a$ and $\beta = b$ we obtain $\phi(t) = c_0 + c_1 t$ for $t \in R$. From each of equations (11) - (12) it follows that $c_1 = 1$, i.e., $\phi(t) = c+t$, $t \in R$, where $c = c_0$. According to (10) we now have $g(\alpha(t)) = \alpha(c+t)$. Setting here $t = \alpha^{-1}(x)$ we have

$$g(x) = \alpha(c+\alpha^{-1})(x)), \quad x \in I,$$

which in view of (9) means that $g = f^c$. This completes the proof.

Using Theorem 2 we can prove the following

Theorem 3. Suppose that $f:I \to I$ is strictly increasing, onto and without fixed points in an open interval I. Let $(f^t)_{t \in R}$ be a measurable iteration group of f such that for each $t \in R$, f^t is continuous in I. Let $(g^t)_{t > 0}$ be a continuous iteration semigroup such that for each $t > 0$, g^t is continuous at least at one point. If

$$f^t \circ g^t = g^t \circ f^t, \quad t > 0, \tag{13}$$

then there is a $c \in R$ such that $g^t = f^{ct}$ for all $t > 0$.

Proof. From (13) we have $f^{nt} \circ g^{mt} = g^{mt} \circ f^{nt}$ for $t > 0$, $n,m \in N$. Setting here $t = s/n$ we get $f^s \circ g^{ws} = g^{ws} \circ f^s$ for all $s > 0$ and all positive rationals w. Take positive reals s and t and a sequence (w_n) of positive rationals such that

$$\lim_{n\to\infty} w_n = \frac{t}{s} .$$

Thus we have

$$f^s \circ g^{w_n s} = g^{w_n s} \circ f^s, \quad n \in N.$$

Since f^s is continous and $(g^t)_{t > 0}$ is a continuous iteration semigroup it follows that

$$f^s \circ g^t = g^t \circ f^s, \quad s,t > 0.$$

In view of Theorem 2 for each $t > 0$ there is a $C(t) \in R$ such that

$$g^t = f^{C(t)} .$$

Now we have

$$f^{C(t+s)} = g^{t+s} = g^t \circ g^s = f^{C(t)+C(s)} , \quad t,s > 0 .$$

Since (f^t) is strictly monotonic with respect to t (cf. formula (9)), it follows that $C(t+s) = C(t)+C(s)$ for $t,s > 0$. Let us take $t_n, t_o > 0$ such that $\lim_{n\to\infty} t_n = t_o$. The continuity of (g^t) with respect to t implies that $\lim_{n\to\infty} g^{t_n} = g^{t_o}$, which means that $\lim_{n\to\infty} f^{C(t_n)} = f^{C(t_o)}$. Making use once more of the strict monotonicity of (f^t) with respect to t, we obtain $\lim_{n\to\infty} C(t_n) = C(t_o)$. Thus $C:(0,\infty) \to R$ is additive and continuous. Consequently, there is a constant $c \in R$ such that $C(t) = ct$ for $t > 0$. This completes the proof.

REFERENCES

[1] Halmos, P.M.: Measure Theory, D. Van Nostrand Company (1950).

[2] Matkowski, J.: On characterization of norms in L^p and functional equations, Proc.Int.Symp. on Functional Equations, Sielpia (Poland), 27.05-02.06 (1984) to appear.

[3] Zdun, M.C.: Continuous and Differentiable Iteration Semigroup, Prace Matematyczne, Silesian University (1979).

Janusz Matkowski

Technical University

Findera 32

PL-43-400 Bielsko-Biała

ON A CRITERION OF ITERATION IN RINGS OF FORMAL POWER SERIES

Günther H. Mehring

During the last years L. Reich and J. Schwaiger studied iterations of automorphisms in formal power series rings.

Before describing the problem we shall introduce some notations. By $\mathbb{C}[[X]] := = \mathbb{C}[[X_1,\dots,X_n]]$ we denote the ring of formal power series with complex coefficients in n indeterminates $X_1,\dots,X_n$. We put these indeterminates together in a column $X := {}^t(X_1,\dots,X_n)$. If $\nu = (\nu_1,\dots,\nu_n) \in \mathbb{N}_0^n$ we use the abbreviations $X^\nu := X_1^{\nu_1}\cdot\ldots\cdot X_n^{\nu_n}$ and $|\nu| = \nu_1 + \dots + \nu_n$. Let $f = \sum_{\nu\in\mathbb{N}_0^n} \tilde{f}_\nu X^\nu \in \mathbb{C}[[X]]$, then we call the smallest number $|\nu|$ for which $\tilde{f}_\nu \neq 0$ the order of f and denote it by O(f). Our main objects will be the automorphisms F of $\mathbb{C}[[X]]$, which are the identity when restricting to the field $\mathbb{C}$ and satisfying the condition $O(F(f)) \geq O(f)$ for all $f \in \mathbb{C}[[X]]$. Such automorphisms are uniquely determined by the image $F(X) \in (\mathbb{C}[[X]])^n$, and we have the following representation: $\mathrm{Aut}(\mathbb{C}[[X]]) \ni F \to F(X) = AX + \sum_{|\nu|\geq 2} f_\nu X^\nu \in (\mathbb{C}[[X]])^n$.

In this formula A means a non-singular complex $n\times n$-matrix and f_ν an element of $\mathbb{C}^n$. The matrix A will be called the linear part lin F of F.

Obviously those automorphisms form a group, denoted by $\mathrm{Aut}(\mathbb{C}[[X]])$. We will say that an automorphism $F \in \mathrm{Aut}(\mathbb{C}[[X]])$ allows an iteration, if there exists a one-parameter family $(F_t)_{t\in\mathbb{C}} \subset \mathrm{Aut}(\mathbb{C}[[X]])$ satisfying the following two conditions:

$$F_{s+t} = F_s \circ F_t \quad \text{for all} \quad t,s \in \mathbb{C} \quad \text{and}$$

$$F_1 = F \; .$$

Using the representation $F_t(X) = A(t)X + \sum_{|\nu|\geq 2} f_\nu(t)X^\nu$ for F_t, we call an iteration of F continuous (resp. analytic), if the functions $t \to f_\nu(t)$ and $t \to A(t)$ are continuous (resp. analytic) for all $\nu \in \mathbb{N}_0^n$, $|\nu| \geq 2$.

If an iteration $(F_t)_{t\in\mathbb{C}}$ of an automorphism F is given, we deduce that the linear parts $\mathrm{lin}\, F_t := A(t)$ form an iteration $(A(t))_{t\in\mathbb{C}}$ of the linear part A of F. This iteration consists of non-singular complex $n \times n$ matrices and its structure is well known (see [3]). Therefore one can describe linear iterations completely.

That is why the main problem of iteration in formal power series rings consists of characterizing those automorphisms F which allow an (continuous resp. analytic) iteration $(F_t)_{t\in\mathbb{C}}$ so that the linear parts $\mathrm{lin}\, F_t$ coincide with a given (continuous resp. analytic) linear iteration $(A(t))_{t\in\mathbb{C}}$ of the linear part lin F of F.

If an automorphism F has a continuous iteration, then it already has an analytic one (see[4]).

Therefore the following theorem gives a complete survey of iteration with regularity conditions (see [6]).

(1) Theorem. Let F be an element of $\text{Aut}(\mathbb{C}[[X]])$ and $(A(t))_{t \in \mathbb{C}}$ be an analytic iteration of lin F.

F allows an analytic iteration $(F_t)_{t \in \mathbb{C}} \subset \text{Aut}(\mathbb{C}[[X]])$ with lin $F_t = A(t)$ for all $t \in \mathbb{C}$ an automorphism $T \in \text{Aut}(\mathbb{C}[X])$ exists, satisfying the following two conditions:

i) $\text{lin}(T \circ F_t \circ T^{-1}) = \overset{r}{\underset{i=1}{\oplus}} \exp(\lambda_i t)\exp(C_i(t))$ for all $t \in \mathbb{C}$. There is $r \leq n$ and the matrices C_i are nilpotent upper triangular matrices fulfilling the conditions $C_i(t+s) = C_i(t) + C_i(s)$, and $C_i(t)C_i(s) = C_i(s)C_i(t)$ for all complex numbers $t,s \in \mathbb{C}$ and $i = 1,\ldots,r$. *)

ii) If $D(t)$ denotes the diagonal matrix which consists of the diagonal elements of $\text{lin}(T \circ F_t \circ T^{-1})$ for all $t \in \mathbb{C}$, the formula $\underset{t \in \mathbb{C}}{\wedge} D(t)X \circ T \circ F \circ T^{-1}(X)$
$= T \circ F \circ T^{-1}(X) \circ D(t)X$

The more general question of characterizing those automorphisms allowing an iteration at all, i.e. an iteration without regularity conditions has been open up till now. An answer will be given by the next theorem.

(2) Theorem. Let be $F \in \text{Aut}(\mathbb{C}[[X]])$ and $(A(t))_{t \in \mathbb{C}}$ an iteration of lin F. F has an iteration $(F_t)_{t \in \mathbb{C}} \subset \text{Aut}(\mathbb{C}[[X]])$ with linear part lin $F_t = A(t)$ for all $t \in \mathbb{C}$, iff an automorphism $T \in \text{Aut}(\mathbb{C}[[X]])$ exists satisfying the following two conditions:

i) $\text{lin}(T \circ F_t \circ T^{-1}) = \overset{r}{\underset{i=1}{\oplus}}\gamma_i(t)\exp(C_i(t))$, for all $t \in \mathbb{C}$. There we have $r \leq n$, and the functions γ_i are homomorphisms of the additive group of $\mathbb{C}$ into the multiplicative group of $\mathbb{C}$ for $i = 1,\ldots,r$. The matrices C_i are nilpotent upper triangular matrices fulfilling the conditions $C_i(t+s) = C_i(t) + C_i(s)$ and $C_i(t)C_i(s) = C_i(s)C_i(t)$ for all $t,s \in \mathbb{C}$ and $i = 1,\ldots,r$.

ii) If $D(t)$ denotes the diagonal matrix consisting of the diagonal elements of $\text{lin}(T \circ F_t \circ T^{-1})$ for all $t \in \mathbb{C}$, then the identity

$$D(t)X \circ T \circ F \circ T^{-1}(X) = T \circ F \circ T^{-1}(X) \circ D(t)X$$

holds for all $t \in \mathbb{C}$.

Before proving the theorem, we should mention two useful facts.

(3) Lemma. The condition ii) of theorem (2) is equivalent to the following: Write
$T \circ F \circ T^{-1}(X) = \text{lin}(T \circ F \circ T^{-1})X + \underset{|\nu| \geq 2}{\sum} g_\nu X^\nu$.

*) These matrices $C_i(\cdot)$ are of the form $C_i(t) = C_i \cdot t$ for all $t \in \mathbb{C}$, because they are analytic.

If $g_{\nu,i} \neq 0$ for $|\nu| \geq 2$ and $1 \leq i \leq n$, then the multiplicative relation

$$\gamma_1^{\nu_1}(t) \ldots \gamma_r^{\nu_n}(t) = \gamma_i(t)$$

holds for all $t \in \mathbb{C}$.

(4) Lemma. Let $(F_t)_{t \in \mathbb{C}}$ be an iteration of F and $T \in \mathrm{Aut}(\mathbb{C}[\![X]\!])$ is mentioned in theorem (2) (resp.(1)), then all automorphisms $T \circ F_t \circ T^{-1}$ fulfill condition ii) in theorem (2) (resp.(1)).

(3) can be shown by simple computation, but (4) is a consequence of a theorem proved by J.Schwaiger (see[6]).

Proof of theorem (2). This criterion is necessary; this was shown by J.Schwaiger (see [6]). It is sufficient, too; in order to prove this we assume that F allows an iteration $(F_t)_{t \in \mathbb{C}}$, and we will deduce some necessary conditions.

Let $F_t(X) = A(t)X + \sum_{|\nu| \geq 2} p_\nu(t) X^\nu$ be an iteration of F(X), satisfying conditions i) and ii) of the theorem for all complex numbers t. Then the translation equation $F_{t+s}(X) = F_t \circ F_s(X)$ holds and leads to the following comparison of the coefficients for $\nu \in \mathbb{N}_0^n$, $|\nu| \geq 2$ and $i = 1,\ldots,n$:

$$\bigwedge_{t,s \in \mathbb{C}} p_{\nu,i}(t+s) = \sum_{j=i}^{n} \gamma_i(t) a_{ij}(t) p_{\nu,j}(s) + \sum_{|\mu|=|\nu|} p_{\mu,i}(t)[(A(s)X)^\mu]_\nu +$$
$$+ \sum_{2 \leq |\mu| < |\nu|} p_{\mu,i}(t) \sum_{\sigma^{(1)}+\ldots+\sigma^{(n)}=\nu} \prod_{j=1}^{n} \sum_{\alpha^{(j,1)}+\ldots+\alpha^{(j,\mu_j)}=\sigma^{(j)}} \prod_{k_j=1}^{\mu_j} p_{\alpha^{(j,k_j)},j}(s) \tag{5}$$

In this formula $p_{\nu,i}$ denotes the ith-component of p_ν and $[(A(s)X)^\mu]_\nu$ the νth-coefficient of $(A(s)X)^\mu$.

Recalling Lemma (3) and setting $p_{\nu,i} = \gamma_i \cdot q_{\nu,i}$ for a suitably chosen function $q_{\nu,i}$, the following identity arises:

$$\bigwedge_{t,s \in \mathbb{C}} q_{\nu,i}(t+s) = \sum_{j=i} a_{ij}(t) q_{\nu,j}(s) + \sum_{|\mu|=|\nu|} q_{\mu,i}(t) [\prod_{j=1}^{n} (\sum_{k=1}^{n} a_{jk}(s) X_k)^{\mu_j}]_\nu +$$
$$+ \sum_{2 \leq |\mu| < |\nu|} q_{\mu,i}(t) \sum_{\sigma^{(1)}+\ldots+\sigma^{(n)}=\nu} \prod_{j=1}^{n} \sum_{\alpha^{(j,1)}+\ldots+\alpha^{(j,\mu_j)}=\sigma^{(j)}} \prod_{k_j=1}^{\mu_j} q_{\alpha^{(j,k_j)},j}(s)$$

From now on we shall abbreviate the last term in (6) by $R_{\nu,i}(\cdot,\cdot)$. Let us define an ordering of $\mathbb{N}_0^n$ as follows:

$$\nu \prec \mu : \bigvee_{i_o \in \{1,\dots,n\}} \nu_{i_o} < \mu_{i_o} \text{ and } \mu_i = \nu_i \text{ for } i = 1,\dots,i_o-1 \,. \tag{7}$$

Now we can deduce a formula which we shall need later on:

$$\prod_{j=1}^{n} \left[\sum_{k=1}^{n} a_{jk}(\cdot)X_k \right]^{\mu_j} = X^{\mu} + \sum_{\substack{|\gamma|=|\mu| \\ \gamma \prec \mu}} e_{\gamma}(\cdot)X^{\gamma} \quad . \tag{8}$$

Using a theorem of Djoković (see [1]) we are able to show that the coefficients $q_{\nu,i}$ are generalized polynomials: If an integer $m \in \mathbb{N}$ ($m = m(q_{\nu,i})$) exists so that $\Delta_s^m q_{\nu,i}$ vanishes for all $s \in \mathbb{C}$, then the above mentioned property is valid. Here Δ denotes the difference operator; it is defined by $\Delta_s h : = h(\cdot + s) - h(\cdot)$ for $s \in \mathbb{C}$ and an ordinary complex function h. If m denotes the smallest integer fulfilling the formula

$$\bigwedge_{s \in \mathbb{C}} \Delta_s^m q_{\nu,i} = 0, \text{ then } q_{\nu,i} \text{ allows the representation } q_{\nu,i} = \sum_{k=1}^{m-1} q^*_{k,(\nu,i)}, \tag{9}$$

where $q^*_{k,(\nu,i)}$ is the diagonalization of the mapping $q_{k,(\nu,i)} : \mathbb{C}^k \to \mathbb{C}$, which is symmetric and additive in each variable.

By simple computation the formula (6) leads to the following:

$$\Delta_s^{m+1} q_{\nu,i}(t) = \sum_{j=i+1}^{n} q_{\nu,j}(s) \, \Delta_s^m a_{i,j}(t) +$$

$$+ \sum_{\substack{|\mu|=|\nu| \\ \mu \succ \nu}} [X^{\mu} + \sum_{\substack{|\gamma|=|\mu| \\ \gamma \prec \mu}} e_{\gamma}(s)X^{\gamma}]_{\nu} \cdot \Delta_s^m q_{\mu,i}(t) + \Delta_s^m R_{\nu,i}(t,s) \quad . \tag{10}$$

With the help of (10) one can verify (9) by a three step induction. We omit the proof, which it is not difficult but rather lengthy and simular to the following considerations.

The coefficients $q_{\nu,i}$ are $\mathbb{Q}$-linear mappings. So the identity $q_{\nu,i}(t) =$

$$= \sum_{k=1}^{m(|\nu|)} q^*_{k,(\nu,i)}(1)t^k \text{ holds for all } \nu \in \{\lambda \in \mathbb{N}_0^n \| \lambda| > 2\}, \ i = 1,\dots,n \text{ and } t \in \mathbb{Q}.$$

The numbers $q^*_{k,(\nu,i)}(1)$ are computable by induction as we shall now. We shall reduce the proof to its essential part.

Let be $\nu \in \mathbb{N}_0^n$, $|\nu| > 2$, and let us assume that all numbers $q^*_{k,(\mu,j)}(1)$ are already known for $\mu \in \{\lambda \in \mathbb{N}_0^n | \ |\lambda| < |\nu|\}$, $\mu \succ \nu$ and $i < j \leq n$. Using the abbreviation

$$b_{\mu,\nu}(.) : = [X^{\mu} + \sum_{\substack{|\gamma|=|\mu| \\ \gamma<\mu}} e_{\gamma}(.)X^{\gamma}]_{\nu},$$

we obtain for $q_{\nu,i}(\cdot)$:

$$\bigwedge_{s,t \in \mathbb{Q}} q_{\nu,i}(t+s) = q_{\nu,i}(t)+q_{\nu,i}(s) + \sum_{k=i+1}^{n} a_{i,k}(t)\cdot q_{\nu,k}(s)$$

$$+ \sum_{\substack{|\mu|=|\nu| \\ \mu\succ\nu}} q_{\mu,i}(t)b_{\mu,\nu}(s) + R_{\nu,i}(t,s) .$$

We can write $\sum_{\chi=1}^{\tilde{m}(A)} e_{\chi}^{(\nu,\mu)}(\cdot)^{\chi}$ for $b_{\mu,\nu}(\cdot)$, and $\sum_{k=1}^{m(A)} d_k^{(j,\sigma)}(\cdot)^k$ for $a_{j,\sigma}(\cdot)$. Since the elements of the matrices $c_i(.)$ are additive functions, hence polynomials over $\mathbb{Q}$. The function $R_{\nu,i}(\cdot,\cdot)$ is obviously a polynomial, so the identity $R_{\nu,i}(t,s) = \sum_{k=1}^{\tilde{N}} \sum_{\phi=1}^{P} h_{k,\phi}^{(\nu,i)} s^{\phi}t^k$ holds for suitably chosen numbers $h_{k,\phi}^{(\nu,i)} \in \mathbb{C}; P,N \in \mathbb{N}_0$ and $s,t \in \mathbb{Q}$.

The above formula is equivalent to the following:

$$\bigwedge_{s,t \in Q} \sum_{k=1}^{\max(m(|\nu|),m(A))} [\sum_{\tau=k+1}^{m(|\nu|)} \binom{\tau}{k} q^*_{\tau,(\nu,i)}(1)s^{\tau-k} -$$

$$\sum_{\sigma=i+1}^{n} \sum_{\chi=1}^{m(|\nu|)} q^*_{\chi,(\nu,\sigma)}(1)d_k^{(i,\sigma)}s^{\chi}]t^k =$$

$$= \sum_{k=1}^{\max(\tilde{N},m(|\nu|))} [\sum_{\substack{|\mu|=|\nu| \\ \mu\succ\nu}} \sum_{\chi=1}^{\tilde{m}(A)} q^*_{k,(\mu,i)}(1)e_{\chi}^{(\nu,\mu)}s^{\chi} + \sum_{\phi=1}^{P} h_{k,\phi}^{(\nu,i)}s^{\phi}]t^k \quad .$$

A two step comparison of coefficients leads to the formula:

$$q^*_{\tau+k,(\nu,i)}(1) = \frac{1}{\binom{\tau+k}{k}} [\sum_{\sigma=i+1}^{n} q^*_{\tau,(\nu,\sigma)}(1)d_k^{(i,\sigma)} + \quad\quad (11)$$

$$+ \sum_{\substack{|\mu|=|\nu| \\ \mu\succ\nu}} q^*_{k,(\mu,i)}(1)e_{\tau}^{(\nu,\mu)} + h_{k\tau}^{(\nu,\mu)}]$$

for $k = 1,\ldots, \max(\tilde{N},m(|\nu|), m(A))$ and $\tau = 1,\ldots, \max(\tilde{P},\tilde{m}(A),m(|\nu|)-k,m(|\nu|))$. We should mention that the above comparisons of coefficients are possible, because the underlying equations are valid for infinite many rationals. Furthermore all steps of proof are equivalent. Thus the proof is nearly complete, if we find complex numbers satisfying the system of equations (11). If this is verified we can ensure the embedding condition by setting $q^*_{1,(\nu,i)}(1) = \frac{1}{\gamma_i(1)}p_{\nu,1}(1)$, because the number $q^*_{1,(\nu,i)}(1)$ is not determined by (11).

Thus we obtain an iteration of F by definition of coefficients as follows:

$$p_{\nu,i}(i) := \begin{cases} \gamma_i(t)\cdot\sum_{k=1}^{m(|\nu|)} q^*_{k,(\nu,i)}(1)t^k , & \text{if } t \in \mathbb{Q} ; \\ & |\nu| \geq 2 ;\; i = 1,\ldots,n \\ 0 , \quad \text{if } t \in \mathbb{C}\setminus\mathbb{Q} ; & \end{cases}$$

The proof comes to the point now: F always allows a natural iteration $(F_n)_{n \in \mathbb{N}}$! Repeating the above considerations for $(F_n)_{n \in \mathbb{N}}$, we get the desired solution of (11), because the comparisons of coefficients still hold.

q.e.d.

The coefficients $q_{\nu,i}$ are generalized polynomials as was shown above. Moreover one can verify that those coefficients are polynomials in additive functions. This can be done by a slight modification of the method of "Complete Linearization", which goes back to a paper of P. Erdös and E. Jabotinsky and was generalized by L. Reich and J. Schwaiger. Therefore we omit the details and refer to the original paper (see [4]).

Let be $(F_t)_{t \in \mathbb{C}}$ an iteration of F and suppose the automorphisms F_t are in normal form, as was described in theorem (2) and lemma (3). Let be $B(F) =$ $= \{F_t \in \mathrm{Aut}(\mathbb{C}[\![X]\!]) \mid t \in \mathbb{C}\}$ and

$$I_i := \{\nu \in \mathbb{N}_0^n \mid \bigwedge_{t \in \mathbb{C}} \gamma_i(t) = \gamma_1(t)^{\nu_1}\cdot\ldots\cdot\gamma_r(t)^{\nu_n}\} \quad \text{for } i = 1,..,r .$$

If $\nu,\mu \in I_i$, we define

$$\nu [\, \mu := \begin{cases} |\nu| < |\mu| & \\ & i = 1,\ldots,r . \\ \nu \,\tilde{<}\, \mu, \quad \text{if } |\nu| = |\mu| & \end{cases}$$

The order $\tilde{<}$ is defined as follows:

$$\nu \,\tilde{<}\, \mu \Leftrightarrow \bigvee_{i_0 \in \{1,\ldots,n\}} \nu_{i_0} > \mu_{i_0} \quad \text{and} \quad \mu_i = \nu_i \quad \text{for } i = 1,\ldots,i_0 - 1 .$$

We verify at once that [is a total ordering of I_i. Furthermore I_i is ordering-isomorphic to N_i, where $N_i = \mathbb{N}$ or $N_i = \{1,\ldots,m(i)\} \subset \mathbb{N}$ means. We have the representation $I_i = \{\beta^{(k)} \in \mathbb{N}_0^n \mid k \in N_i\}$ satisfying the condition: $\beta^{(k)} [\, \beta^{(l)}$ for $k < l$.

Let us denote by M_{N_i,N_i} the set

$$\{\phi : N_i \times N_i \to \mathbb{C} \mid \bigwedge_{j \in N_i} \bigvee_{M(j) \in N_i} \bigwedge_{K \geq M(j)} \phi(k,j) = 0\}$$

for $i = 1,\ldots,r$.

M_{N_i,N_i} becomes a $\mathbb{C}$-algebra by pointwise definition of the linear operations and by definition of multiplication as follows:

$$(\phi.\psi)(j,l) := \sum_{k \in N_i} \phi(j,k).\psi(k,l) \text{ for } j,l \in N_i \text{ and } \psi,\phi \in M_{N_i,N_i} .$$

Now we are able to define the "linearization mapping":

$$L : \mathcal{B}(F) \ni G \to \bigoplus_{i=1}^{r} ([G(X)^{\beta(k)}]_{\beta(l)})_{k,l \in N_i} \in \bigoplus_{i=1}^{r} M_{N_i,N_i} .$$

Let us briefly write $L_i(G)$ for $([G(X)^{\beta(k)}]_{\beta(l)})_{k,l \in N_i}$. The next theorem gives a description of the linearization mapping as detailed as necessary:

(12) Theorem. On the assumptions mentioned above, L has the following properties:

i) L is injective.

ii) $L_i(G \circ H) = L_i(G) \cdot L_i(H)$ for $G,H \in \mathcal{B}(F)$ and $i = 1,\ldots,r$.

iii) The coefficients of G_j appear in the matrix $L_i(G)$ for suitably chosen $j \in \{1,\ldots,n\}$, $i = 1,\ldots r$.

iv) Let $(F_t)_{t \in \mathbb{C}}$ be an iteration of F, then $(L_i(F_t))_{t \in \mathbb{C}}$ is an iteration of $L_i(F)$ which consists of upper triangular matrices for $i = 1,\ldots,r$.

With the help of a theorem of McKiernan (see [2]) one can deduce that the iteration $(L_i(F_t)_{t \in \mathbb{C}}$ of $L_i(F)$ allows the representation:

$$\bigwedge_{t \in \mathbb{C}} \frac{1}{\gamma_i(t)} L_i(F_t) = \sum_{\nu=0}^{\infty} \frac{1}{\nu!} (\mathcal{C}_i(t))^{\nu}, \quad i = 1,\ldots,r . \tag{13}$$

The family $(\mathcal{C}_i(t))_{t \in \mathbb{C}}$ also of upper triangular matrices, but the diagonal elements vanish for all $t \in \mathbb{C}$. Moreover $(\mathcal{C}_i(t))_{t \in \mathbb{C}}$ satisfies the conditions

$$\left.\begin{aligned} \mathcal{C}_i(t+s) &= \mathcal{C}_i(t) + \mathcal{C}_i(s) \\ \mathcal{C}_i(t)\mathcal{C}_i(s) &= \mathcal{C}_i(s)\mathcal{C}_i(t) \end{aligned}\right\} \quad \text{for all } t,s \in \mathbb{C} .$$

Hence by (12) and (13) the coefficients $q_{\nu,i}$ are polynomials in additive functions.–

I wish to express my gratitude to L. Reich and J. Schwaiger for many discussions and helpful comments while this paper was written.

REFERENCES

[1] Djoković, D.Z.: A representation theorem for $(X_1-1)(X_2-1)\ldots(X_n-1)$ and its application; Annales Polonici Math. XXII (1969).

[2] Mc Kiernan, M.A.: The matrix equation $a(x \circ y) = a(x)+a(x)a(y)+a(y)$; Aequationes Mathematicae 15 (1977).

[3] Reich, L.: Über analytische Iteration linearer und kontrahierender biholomorpher Abbildungen; Ber. d. Ges. f. Math. u. Datenverarbeitung, Nr.42 (1971).

[4] Reich, L., Schwaiger, J.: Linearisierung formal-biholomorpher Abbildungen und Iterationsprobleme; Aequationes Mathematicae 20, p.244 (1980).

[5] Reich, L., Schwaiger, J.: Über einen Satz von Shl.Sternberg in der Theorie der analytischen Iteration; Monatshefte für Math. 83, p.207 (1977).

[6] Schwaiger, J.: Normal Forms for Systems of Formal Power Series Commuting in Pairs and Iteration Problems; In this volume.

Günther H. Mehring
Institut für Mathematik
Universität Graz
Brandhofgasse 18
A-8010 Graz, Austria

ROTATION SEQUENCES AND BIFURCATIONS STRUCTURE OF ONE-DIMENSIONAL ENDOMORPHISMS

Christian Mira

1. INTRODUCTION

Let T be a one-dimensional quadratic endomorphism which can always be reduced by a linear change of variable (cf.[1c] and [3] p.97) to the form

$$x_{n+1} = x_n^2 - \lambda, \quad n = 0,1,2,... \tag{1}$$

where x is a real variable, λ a real parameter. A cycle of order k, k = 2,3,..., is constituted by k points, given by the solutions of $x = T^k x$, $x \neq T^l x$, $l < k$. Every x of such a solution defines k-1 consequents obtained by means of (1), which are also roots of $x = T^k x$, $x \neq T^l x$. The k-th consequent of any root coincides with itself, and the corresponding set of k points forms a cycle of order k. A cycle of order k = 1 is a fixed point of T.

Since Myrberg [1] it is known that the number N_k of all possible cycles having the same order k, and the number $N_\lambda(k)$ of bifurcation values giving rise to these cycles, increase very rapidly with k (k = 5, $N_k = 6$, $N_\lambda(k) = 3$; k = 10, $N_k = 99$, $N_\lambda(k) = 51$; k = 30, $N_k = 35790267$, $N_\lambda(k) = 17895679$) (cf.[1c] and [3] p.104). These cycles with the same order k differ from one another by the type of cyclic transfer of one of their points by k successive iterations by T.

Myrberg [1c] was the first, in 1963, to characterize this type of cyclic transfer, by a sequence of (k-2) signs + and -, when one of the k points of the cycle is located at the extremum x = 0 of T. Such a characterization constitutes a *rotation sequence* [r] *with a binary base*, which defines also the λ value, λ^o, corresponding to the considered cycle, λ^o being a solution of a radical equation [1c]. Implicitly the Myrberg papers [1,2] give an extension of the rotation sequence notion to the limit cases $k \to \infty$ corresponding to an accumulation λ^o_{lim} of λ^o. The ordering laws of such sequences are given in the Myrberg's paper [1c]. These laws were rediscovered in 1980 [4] with the same binary code (R for - and L for +, (k-1) symbols RL in [5] but the first one, always the same R(-), is omitted in the Myrberg's representation).

Another type of characterization was proposed in [3,6], by defining a *rotation sequence* [u] *with a decimal base*, independant of the position of the cycle points with respect to the extremum of T. It has been shown in [3,6] that such rotation sequences have fundamental properties directly related to the fractal bifurcations structure of T, called "*box-within-a-box*" structure in [3,6-9] (where a description can be found), or "*embedded box*" in [11]. Another definition [g] of a rotation sequence with a decimal base was given in [12,13], but with the constraint that one of the points of the cycle must be located at the extremum of T. Papers [12,13] give the properties of [g], and show that the correspondance between [u] and [g] is a bijection.

The purpose of this paper is to give an amplified presentation of the properties of the rotation sequences [u] with respect to that first exhibited in [6] (1978) and [3] (1980), specially in relation with the "box-within-a-box" bifurcations structure of T. So §2, 3 recall notions and results of [1,3] but properties (P_1) (P_2) and §4-6 correspond to unpublished results.

2. DEFINITIONS OF THE "BINARY" AND "DECIMAL" ROTATION SEQUENCES [r] and [u]

Let T' be now the one-dimensional endomorphism:

$$x_{n+1} = f(x_n,\lambda),\ n = 0,1,2,...,\quad \text{or}\quad x_{n+1} = T'x_n \tag{2}$$

where $f(x,\lambda)$ is a continuous function of the real variable x, and the real parameter λ, $f(x,\lambda)$ having only one extremum for $x = x_e$. Furthermore, $f(x,\lambda)$ is such that T'^k cannot intersect $x_{n+1} = x_n$ in more than 2^k points.

The "binary" rotation sequence of a cycle of order k, having one of its points in x_e, is constituted by (k-2) signs associated with the set of the consequents of x_e having the rank i, i = 2,3,..., k-2.

Let us consider a cycle of order k, having one of its points in x_e (i.e. the eigenvalue, or multiplier S = 0), and the (k-2) points $T'^i x_e$, i = 2,3,..., k-2. With each of these consecutive points is associated the sign + if $T'^i x_e > (<) x_e$ when x_e is a minimum (maximum) for $f(x,\lambda)$, and the sign - in the contrary case. The set of the (k-2) signs, so obtained from i = 2 to k, is the "binary" rotation sequence [r], or Myrberg's rotation sequence of this cycle [1c].

Let us now consider a cycle of order k for T' without any restrictive condition on the position of its points. If $f(x,\lambda)$ has a minimum (maximum), the k points of cycle are numbered from 1 to k with increasing (decreasing) abscissae. The "decimal" rotation sequence of this cycle is the set [u] of the k numbers, $[u] = [1,k,\alpha_3,\alpha_4,...\alpha_k]$, $\alpha_1 \equiv 1$, $\alpha_2 \equiv k$, the α_i, i = 1,2,...,k, being the running numbers of the points obtained by successive applications of T', beginning by the point numbered 1. This sequence [u] characterizes the exchange of the points of a given cycle, and does not change when λ varies. It is an invariant associated with the considered cycle. Two cycles of order k, born from the same bifurcation, have the same [u].

The sequences [u] and [r] are in relation via:

(P1) - The Myrberg's rotation sequence [r] is obtained from the "decimal" rotation sequence [u], by omitting 1 and α_k into [u], by replacing the $\alpha_i < \alpha_k$ by the sign -, and the $\alpha_i > \alpha_k$ by the sign + [10].

We have also:

(P2) - Let [p] be a permutation of the k first integers. On the line $x_{n+1} = x_n$ of the (x_n, x_{n+1}) plane, k points are considered, numbered from 1 to k with increasing (decreasing) abscissae when $f(x,\lambda)$ has a minimum (maximum). Following the order given by [p], the k points are joined by passing through each point a parallel line to the x_n-axis, and a parallel line to the x_{n+1}-axis. The last point of p is also joined to the first similarly. So is obtained a closed curve, constituted by segments of straight lines, with k angular points p_i, $i = 1,2,\ldots,k$, outside $x_{n+1} = x_n$. The necessary and sufficient condition for [p] being a rotation sequence [u], associated with a cycle of (2), is that it should be possible to pass through the k points p_i a continuous curve with only one extremum cf. (Fig.1,2).

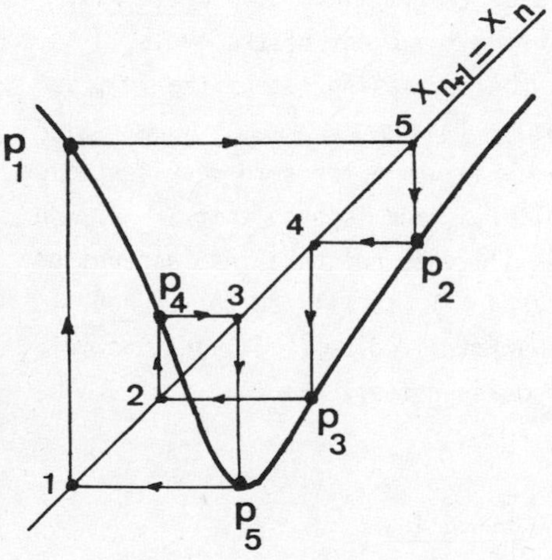

Fig. 1

[u] = [1 5 4 2 3] is a rotation sequence for a cycle of T'. It is possible to pass an infinity of continuous curves with only one extremum by the points p_i, $i = 1,\ldots,5$.

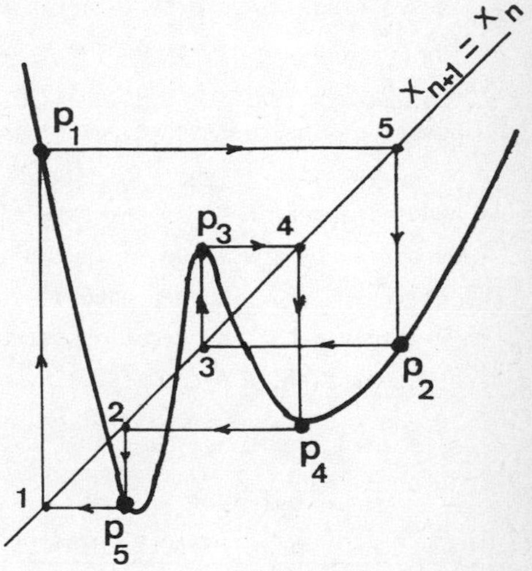

Fig. 2

[u] = [1 5 3 4 2] is not a rotation sequence for a cycle of T'. It is impossible to pass a continuous curve with only one extremum by the points p_i.

3. FIRST NUMBERING OF THE CYCLES. NON EMBEDDED REPRESENTATION

The cycles numbering is based on Myrberg's ordering law stated for the rotation sequences [r][1c]. A more "compact" statement of this law is given in [4].

(P3) - Let [r] = (A)(B),[r'] = (A)(B') be two Myrberg's rotation sequences, having in common the same j first signs represented by (A); (B) (B') being two different sequences of signs +, -. Let b_1 be the first sign of (B), and b_1' the first sign of (B'), this sign being replaced by 0 if [r], or [r'] is limited to (A). The following ordering between b_1, b_1' is introduced: (-) < 0 < (+). Then it is said that [r'] > [r] if: a) (A) contains an even number, or zero number, of signs -, with $b_1' > b_1$. b) (A) contains an odd number of signs -, with $b_1' < b_1$.

For example: (++++) > (++++-), (++--) < (++--+), (+--+-) > (+---+-), (+---+-) < < (+----). Considering now (1), $f(x,\lambda) \equiv x^2-\lambda$, from [1c] it appears that each [r] corresponds to $\lambda = \lambda^0$, giving a point located at $x_e = 0$ for the cycle defined by [r]. Then the ordering law (P3) gives all the inequality relations between the λ^0. It must be observed that (P3) concerns the cases when [r] is a finite, or an infinite set of signs, as it appears in the Myrberg's papers [1c,2] (cf.also [10]).

Myrberg's ordering law permits to arrange the cycles with the same order k in an increasing set. Now, whatever be the eigenvalue (multiplier) S, a cycle will be represented by the double symbol (k;j'), k being the order of the cycle, j' being the number of the corresponding [r] into the increasing set of the [r], defined for S = 0 and associated with all the cycles having the same order k. Two cycles of order k appearing from the same bifurcation have the same representation (k;j'). They differ by the value of the multiplier S, one is such that S < 1, the other is such that S > 1. The representation (k;j') does not take into account the fractal "box-within-a-box" bifurcations structure of T. It will be called non embedded representation. A cycle is entirely characterized by (k;j') ant its multiplier S. Using (P1), (P3) permits also the ordering of all the [u].

4. DECOMPOSABLE AND INDECOMPOSABLE ROTATION SEQUENCES [u]

The disposition of the components of a decimal rotation sequence [u] is directly related with the box-within-a-box bifurcations structure of T. This fundamental property of the [u] was described for the first time in [6], but was used before to number the boxes of the structure [3,7,9]. It is shown in [6] that, for a subset of the rotation sequences, it is possible to break down [u] into partial sequences (cf.[3] p.128 and [6]). For breakdown operations, it seems that the form [u] is more convenient to manipulate than the other above mentioned decimal form [g] which uses the "evolution matrix" tool [13].

Definition. *Let* $[u] = [1k\alpha_3 \ldots \alpha_k]$ *be the rotation sequence associated to the cycle* $(k;j)$.*Let* β *one of the divisors of* k *(if it exists) such that the partial sequence* $[u_{k/\beta}] = [1\alpha_{\beta+1}\alpha_{2\beta+1} \ldots \alpha_{p\beta+1}]$, *of order* $m = k/\beta$, $p = m-1$, *extracted from* $[u]$ *is a rotation sequence (cf.P2). Let* $[u_\beta]$ *(remainder) be the partial sequence of order* β, *which is obtained by forming* β *sets of* k/β *consecutive points, the* k *points having the numbering used to define* $[u]$. *Similarly, the* β *sets are numbered from* 1 *to* β. *The sequence* $[u_\beta]$ *is written from* $[u]$ *by following the exchange order of the sets with successive applications of* T'. *If, as* $[u_{k/\beta}]$, $[u_\beta]$ *is a rotation sequence,* $[u]$ *is said decomposable; then* $[u_{k/\beta}]$ *and* $[u_\beta]$ *are said the components factors, or the components of* $[u]$. *In the opposite case,* $[u]$ *is said indecomposable.*

Example. Let $[u] = [1,15,9,4,12,3,13,7,6,10,2,14,8,5,11]$. Among the divisors of $k = 15$, only $\beta = 5$ gives $[u_{15/5}] = [1,3,2]$ which verifies (P2) (corresponding cycle $k = 3$; $j' = 1$). Let $g_1 = (1,2,3)$, $g_2 = (4,5,6)$, $g_3 = (7,8,9)$, $g_4 = (10,11,12)$, $g_5 = (13,14,15)$ be the $\beta = 5$ sets of points. Taking into account $[u]$, the exchange order of these sets is $[u_5] = [g_1 g_5 g_3 g_2 g_4] = [1\ 5\ 3\ 2\ 4]$, rotation sequence of the cycle ($k = 5$; $j' = 1$). From this step, $[u_{15/3}]$ is indecomposable. Afterwards, β will be said the *decomposing factor* and $[u_{k/\beta}]$ the *quotient*.

Remarks. a) If k is a prime number, $[u]$ is always indecomposable. Nevertheless, $[u]$ can be indecomposable with k different from a prime number. So $[1\ 6\ 5\ 3\ 2\ 4]$, ($k = 6$; $j' = 3$) is indecomposable, but $[1\ 6\ 3\ 4\ 2\ 5]$ associated with the cycle ($k = 6$; $j' = 1$) is decomposable with $\beta = 2$ giving $[u_{6/2}] = [1\ 3\ 2]$. b) Let us consider the Myrberg's bifurcations chain (period, or order, doubling for a cycle) [1c]:

$$\text{attr.cycle } k2^i \rightarrow \text{rep.cycle } k2^i + \text{attr.cycle } k2^{i+1} \tag{3}$$

"attr.(rep.) cycle $k2^i$" representing "attractive (repulsive) cycle of order $k2^i$", $k = 1,3,4,5,\ldots$, $i = 0,1,2,\ldots$ With $k = 1$, a cycle or order 2^i born from (3) is such that $[u] = [u]_{2^i} = [1, \gamma-1, \alpha_1, \gamma-\alpha_1, \alpha_2, \gamma-\alpha_2, \ldots \alpha_{m/2-1}, \gamma-\alpha_{m/2-1}]$ $\gamma = 2^i+1$, $m = 2^i$, $\alpha_1 = 2^{i-1}$, and $[u]_{2^{i-1}} = [1, \alpha_1, \alpha_2, \ldots, \alpha_{m/2-1}]$. Applying i times the decomposing factor $\beta = 2$, or once the decomposing factor $\beta = 2^i$, we obtain the indecomposable quotient $[u] = 1$.

5. SECOND NUMBERING OF THE CYCLES. EMBEDDED REPRESENTATION

Let U_I the set of the indecomposable rotation sequence $[u]$. With (P1) and (P3), it is possible to arrange each set of the indecomposable $[u]$, having the same order k, in an increasing set of elements. Let j be the number of such a $[u]$ into this set. A cycle corresponding to $[u] \in U_I$ will be represented by $(k;j)$. For example, $[u] = [1\ 6\ 5\ 3\ 2\ 4]$ corresponds to ($k = 6$; $j' = 3$) with the first numbering, and to (k = 6; j = 1) with the second numbering, [u] being the first indecomposable rotation sequence in the increasing set of the indecomposable [u] of order k = 6.

The embedded representation appears in the following situations when [u] is decomposable with $[u] \neq [u]_{2^i}$.

5.1. $[u_{k/\beta}]$ and $[u_\beta]$ are indecomposable.

Let $\beta = k_1$, $k/\beta = k_2$. The rotation sequence is only once decomposable with the decomposing factor k_1. To $[u_\beta]$ and to $[u_{k/\beta}]$, which are indecomposable, the indices j_1, j_2 are respectively associated, as indicated above. Then [u] is the rotation sequence of a cycle, which will be represented by $(k_1.k_2;j_1,j_2)$. For the example of §4 with k = 15, $k_1 = 5$, $k_2 = 3$, $j_1 = j_2 = 1$. It is possible to write [u] = $= [u_{k_1}^{j1}] \circ [u_{k_2}^{j2}]$ or more compactly $[u] = [u_{k_1.k_2}^{j1,j_2}]$

5.2. $[u_{k/\beta}]$ is indecomposable and $[u_\beta] \equiv [u]_{2^i}$.

Let $k/\beta = k_2$. Then [u] is the rotation sequence of a cycle, which will be represented by $(2^i k_2;1,j_2)$. Such a cycle appears into a "box", of the fractal bifurcations structure, called Ω_{2^i} in [3,7-10], at the boundary of the box $\Omega_{2^i.k_2}^{1,j_2}$

5.3 $[u_\beta]$ is indecomposable, and $[u_{k/\beta}] \equiv [u]_{2^i}$ $(\gamma = k/\beta + 1)$.

Let $\beta = k_1$. The corresponding cycle is born from (3) with $k = k_1 2^i$. This cycle will be represented by $(k_1 2^i;j_1)$. So [u] = [1,12,8,4,9,5,2,11,7,3,10,6] is the rotation sequence of the cycle $(3.2^2;1)$, since $[u_{12/3}]$ = [1 4 2 3] is the rotation sequence of the cycle having the order 2^2 (m = 2).

5.4 General case, $[u_{k/\beta}]$ and (or) $[u_\beta]$ are decomposable.

Among the set of all the divisors of k, all the decomposing factors of [u] are extracted. They appear in the form of products: $k_1, k_1.k_2, \ldots, k_1.k_2 \ldots k_{a-1}$, and we have $k = k_1.k_2 \ldots k_{a-1} k_a$. The smallest k_1 gives the first indecomposable component of [u], represented by $[u_{k_1}^{j_1}]$, the index j_1 being the corresponding number in the increasing set of the indecomposable [u], of order k_1. After the quotient $[u_{k/k_1}]$ is considered, with the corresponding factors $k_2, k_2.k_3, \ldots, k_2 k_3 \ldots k_{a-1}$. The smallest is k_2 giving the 2nd indecomposable component $[u_{k_2}^{j_2}]$ with the index j_2.

Continuing till rank (a-1), the last decomposing factor k_{a-1}, the index j_{a-1}, and the component $[u_{k_{a-1}}^{j_{a-1}}]$ are defined. The quotient of the decomposition is $[u_{k/k_1/\ldots}$ $\ldots/k_{a-1}]$, which is now either indecomposable with the order k_a and the index j_a, or having the form $[u]_{2^i}$.

In the first situation, [u] is the rotation sequence of a cycle represented by $(k_1.k_2\ldots k_a;j_1,j_2,\ldots,j_a)$, and [u] is written in the form $[u] = [u_{k_1}^{j_1}] \circ [u_{k_2}^{j_2}] \circ \ldots$ $\ldots \circ [u_{k_a}^{j_a}]$.

In the second situation, [u] is the rotation sequence of a cycle born from (3), and represented by $(k_1.k_2\ldots k_{a-1}2^i;j_1,j_2,\ldots,j_{a-1})$.

In the described decomposition, if one of the factors is $k_n = 2^l$, n = 1,2,... ...,a-1, l = integer, and such that $[u_{k_n}] \equiv [u]_{2^l}$, it is not decomposed for obtaining the smallest decomposing factor. To $[u_{k_n}]$ the index $j_n = 1$ is associated, and it is proceeded as in §5.2. Then the corresponding cycle is represented by $(k_1\ldots k_{n-1}2^l k_{n+1}\ldots k_a;j_a,\ldots,j_{n-1},1,j_{n+1},\ldots,j_a)$, or by $(k_1\ldots k_{n-1}2^l k_{n+1}\ldots k_{a-1}2^i;$ $j_a,\ldots,j_{n-1},1,j_{n+1},\ldots,j_{a-1})$ in the second above mentioned situation.

So [u] = [1,24,16,8,17,9,4,21,13,5,20,12,3,22,14,6,19,11,2,23,15,7,18,10] is the rotation sequence of the cycle $(3.2^1.4;1,1,1)$, and [u] = [1,24,16,8,17,9,6,19, 11,4,21,13,2,23,15,7,18,10,5,20,12,3,22,14] is the rotation sequence of the cycle $(3.4.2^1;1,1)$.

Remarks. The operations of §5 constitute a decomposition of [u] into a set of indecomposable rotation sequences. So it is obtained a representation of all the cycles of (1). (P2) shows that this decomposition is unique. Moreover, the described operations indicate that the representation $(k_1\ldots k_a;j_1,\ldots,j_a)$ is not commutative.

6. THE COMPOSITION OF INDECOMPOSABLE ROTATION SEQUENCES

It is the inverse problem of the problem treated in §5. Composing two rotation sequences $[u_{k_1}^{j_1}]$, $[u_{k_2}^{j_2}] \in U_I$ amounts to define the arrangement of the rotation sequence components associated to the cycle $(k_1k_2;j_1j_2)$, with $[u_{k_1.k_2}^{j_1,j_2}] = [u_{k_1}^{j_1}] \circ$ $\circ\ [u_{k_2}^{j_2}]$.

The solution of this problem must take into account the representation of the curve T^{k_1} in the (x_n,x_{n+1}) plane, for the λ values just after the birth of the cycle $(k_1k_2;j_1,j_2)$. For these λ values, k_1 segments (C_l) of T^{k_1}, $l = 1,\ldots k_1$, with a parabolic aspect, have a projection on the x_n-axis, each one containing a set of k_2 points of the cycle $(k_1k_2;j_1,j_2)$. These k_2 points have a cyclic transfer defined by the index j_2, by successive applications of T^{k_1}. The k_2 points of a set are numbered with increasing abscissae if the extremum of the corresponding C_l is a minimum, and with decreasing abscissae if it is a maximum. The rotation sequence

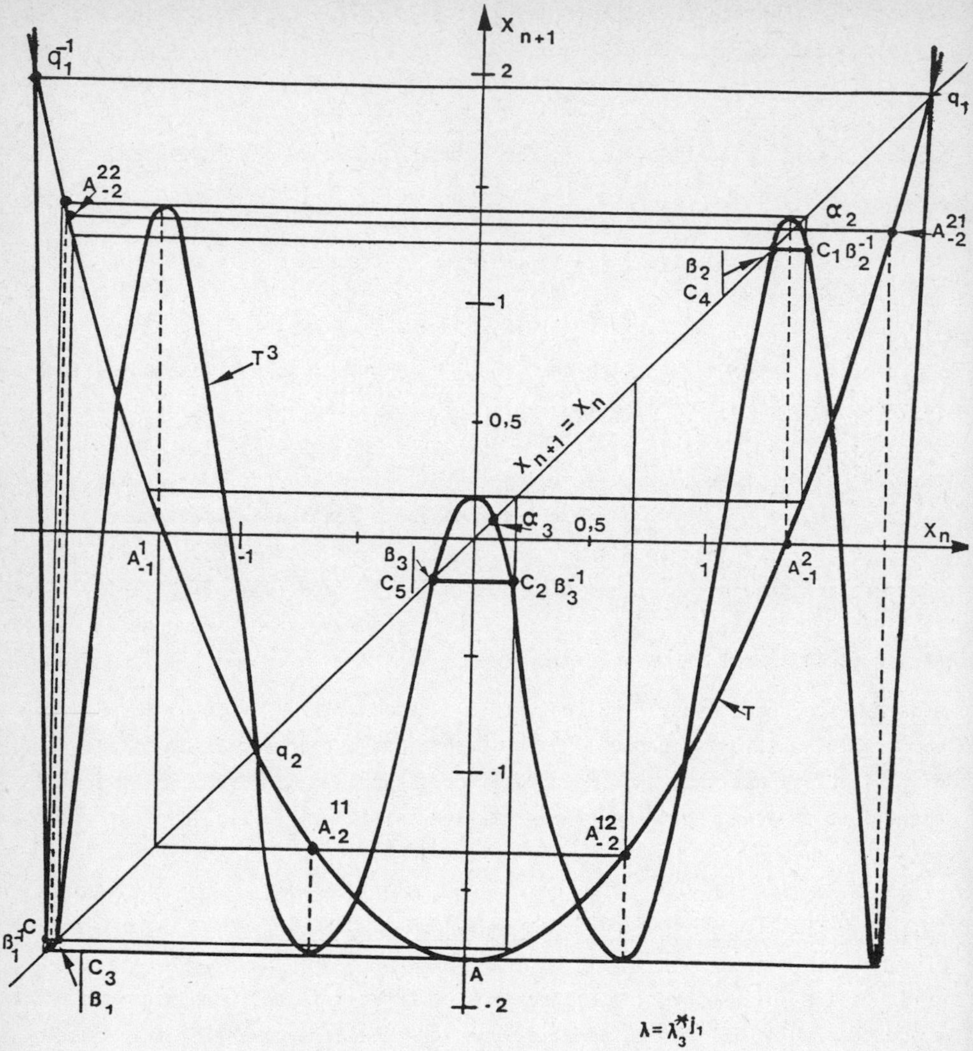

Fig. 3.

α_i, β_j, i = 1,2,3, are respectively the 3 points of the cycle k = 3 with S < 1, and of the cycle k = 3 with S > 1. C_j, j = 0,1,...,5, $C_0 \equiv C$, is the consequent of rank j of the minimum A. The projections of the parabolic segments (C_l), l = 1,2,3, of T^3 on x_n-axis are respectively $\overline{CC_3}, \overline{C_2C_5}, \overline{C_1C_4}$. (C_1) has a minimum, (C_2), (C_3) have each one a maximum.

$[u^{j_1,j_2}_{k_1k_2}]$ is then directly obtained from $[u^{j_1}_{k_1}]$ and $[u^{j_2}_{k_2}]$, as it is shown from the following example.

Example. Let $[u^1_3] = [1\ 3\ 2]$ and $[u^1_4] = [1\ 4\ 3\ 2]$; $[u^{1,1}_{3.4}] = [u^1_3] \circ [u^1_4]$ must be constructed. With increasing abscissae, the first segment (C_1) of T^3 has a minimum, $(C_2),(C_3)$ having a maximum (cf.Fig.3). Each of (C_l), $l = 1,2,3$, contains a set g_l of $k_2 = 4$ points with a cyclic transfer, by application of T^3, defined by $[u^1_4]$. We have $g_1 = (1,2,3,4)$, $g_2 = (5,6,7,8)$, $g_3 = (9,10,11,12)$ by numbering with increasing abscissae. By T^3, the cyclic transfer into g_1 is made with the order [1 4 3 2], (C_1) having a minimum. For g_2, g_3, the extremum of which being a maximum, the cyclic transfer [1 4 3 2] must be considered by numbering with decreasing abscissae the points of g_2, g_3. So we consider $\bar{g}_2 = (8,7,6,5)$ with the cyclic transfer given by [8 5 6 7], and $\bar{g}_3 = (12,11,10,9)$ with [12,9,10,11]. Writing $[u^1_3] = [g_1,g_3,g_2]$ and following the points of each set with the indicated order, we have $[u^{1,1}_{3,4}] = [1,12,8,4,9,5,3,10,6,2,11,7]$.

The general case, with more than two indecomposable rotation sequences, is treated similarly by considering the compositions with successive pairs of [u] in the inverse order of §5.4.

7. CONCLUSION

Each of the three representations of rotation sequences has its own interest. The binary form [r] is more practical when using the ordering law, and for the representation of the limit cases with $k \to \infty$ [10]. The decimal form [u] is particularly convenient for the operations of decomposition, as above mentioned. These operations are directly related with the fractal bifurcations structure, of "box-within-a-box" type, and have given the symbolic representation of the boxes of this structure [3,6-10]. As for the other decimal form [g] (in bijection with [u]) introduced in [12,13], it seems suitable for the treatment of the problems of topological entropy.

REFERENCES

[1] Myrberg, P.J.: Iteration der reellen Polynome zweiten Grades I, II, III. Ann.Acad.Sci.Fenn., Ser.A(a) 256, p.1-10 (1958). (b) 268, p.1-13 (1959), (c) 336, p.1-8 (1963).

[2] Myrberg, P.J.: Iteration von Quadratwurzeloperationen. Ann.Acad.Sci.Fenn., Ser.A, 259, p.1-10 (1958).

[3] Gumowski, I., Mira, C.: Dynamique Chaotique. Cepadues Editions, Toulouse, (1980).

[4] Collet, P., Eckman, J.P.: Iterated Maps on the Interval as Dynamical Systems. Birkhauser, 1980.

[5] Metropolis, N., Stein, M.L.: On finite limit sets for transformations on the unit interval. Journal of Combin.Th.(A), 15, p.25-44 (1973).

[6] Mira, C.: Systèmes à dynamique complexe et bifurcations. RAIRO Automatique, 12, no.1, p.63-94 (1978); 12, no.2, p.171-190 (1978).

[7] Gumowski, I., Mira, C.: Accumulations de bifurcations dans une récurrence. C.R.Acad.Sc.Paris (A), 281, p.45-48 (1975).

[8] Mira, C.: Accumulations de bifurcations et "structures boites emboiteés dans les recurrences et transformations ponctuelles. Proceedings of ICNO 7th Berlin 1975, Akademic Verlag, Berlin, p.81-93 (1977).

[9] Mira, C.: Actes des Journées Analyse Non Linéaire, Application à la Mécanique (mai 1983), IRMA, Lille, vol.5, fasc.2, p.M 1-11.

[10] Mira, C.: Coexistence de structures de bifurcations boites emboitées et boites en files dans un endomorphisme uni-dimensionnel. Publication "Syst.Dyn.", INSA 83-2 (1983).

[11] Guckenheimer, J.: Bifurcations theory and applications in scientific disciplines. Ann. of N.Y.Acad. of Sc.vol. 316, p.78-85 (1979).

[12] Gillot, C.: Caractérisation des suites de rotation d'un endomorphisme unimodal de (0,1). C.R.Acad.Sc.Paris (1), 293, (1981), p.249-252.

[13] Gillot, C.: Structures des itinéraires symboliques et de l'entropie topoloque des transformations unimodales. Thèse de Doctorat ès-Sciences Mathêmathiques, Université Paul Sabatier, Toulouse, no. 1134, 15 mars 1984.

Christian Mira
I.N.S.A.
Avenue de Rangueil
F-31077 Toulouse Cedex

CHAOS ALMOST EVERYWHERE

Michał Misiurewicz

According to the definition by Li and Yorke [1], a continuous map $f : I \to I$ (where I is an interval) is chaotic if there are infinitely many periodic points of different periods and there exists an uncountable set $S \subset I$ with the following properties:

(1) for every $x \in S$ and y periodic,

$$\limsup_{n\to\infty} |f^n(x) - f^n(y)| > 0 ,$$

(2) for every $x,y \in S$ with $x \neq y$,

$$\limsup_{n\to\infty} |f^n(x) - f^n(y)| > 0 ,$$

(3) for every $x,y \in S$,

$$\liminf_{n\to\infty} |f^n(x) - f^n(y)| = 0 .$$

Such a set S is called a scrambled set.

Smítal gave examples of a scrambled set of full outer Lebesgue measure [2] and of positive Lebesgue measure [3]. Here we prove a stronger result:

Theorem. There exists a continuous map of the interval [0,1] onto itself for which there exists a scrambled set of Lebesgue measure 1.

In other words, this map is chaotic almost everywhere in a very strong sense (although this is still a topological kind of chaos; in (1) - (3) statements are about all points of S).

Since this map is highly non-differentiable, one can ask for instance whether such examples of class C^r ($r = 1,2,\dots$) can be constructed.

The idea of the proof is as follows. For a standard "tent" map we construct (using symbolic dynamics) a sequence of scrambled sets. Their union also is a scrambled set and is dense. Each of them supports a non-atomic measure. A weighted average of these measures is a probabilistic non-atomic measure, positive on non-empty open sets. Then we transport everything (the map, the scrambled set) by the homeomorphism which sends this measure to the Lebesgue measure.

Now we begin the detailed proof. Denote:

- by I the interval [0,1],
- by J the interval [0,1),

- by f the "tent" map from I onto itself, namely

$f(x) = 1 - |2x - 1|$,

- by X the space of all 0-1 sequences, $X = \{0,1^{\mathbb{N}}\}$ ($\mathbb{N}$ is the set of all non-negative integers),
- by h the shift on X.

It is clear that for every $\underline{x} = (x_n)_{n=0}^{\infty} \in X$ there exists a unique point $p(\underline{x}) \in I$ such that for all $n \in \mathbb{N}$,

$$f^n(p(\underline{x})) \in \begin{cases} [0,1/2] & \text{if} \quad x_n = 0 \\ [1/2,1] & \text{if} \quad x_n = 1 \end{cases}$$

(e.g. $1/2 = p(0,1,0,0,\ldots) = p(1,1,0,0,\ldots)$).
The map $p : X \to I$ defined in such a way is continuous and $p \circ h = f \circ p$.

For $\underline{x} = (x_n)_{n=0}^{\infty} \in X$ and $A \subset \mathbb{N}$ such that $\mathbb{N} \setminus A$ is infinite, we define $F(A,\underline{x}) = (y_n)_{n=0}^{\infty} \in X$ as follows: we take the unique bijection $b : \mathbb{N} A \to \mathbb{N}$ preserving order and set

$$y_n = \begin{cases} 1 & \text{if} \quad n \in A , \\ x_{b(n)} & \text{if} \quad n \notin A . \end{cases}$$

Fix an irrational number α. For $t \in J$ and $n \in \mathbb{N}$ we define

$$a(t,n) = \begin{cases} 0 & \text{if} \quad \{t+n\alpha\} \in [0,1/2) \\ 1 & \text{if} \quad \{t+n\alpha\} \in [1/2,1) , \end{cases}$$

(where $\{s\} = s - E(s)$ is the fractional part of s) and then the map $a : J \to X$ by $a(t) = (a(t,n))_{n=0}^{\infty} \in X$. In other words, $a(t)$ is the code of the point t, corresponding to the map $R_\alpha : t \to t + \alpha \pmod 1$ and the partition of J into $[0,1/2)$ and $[1/2,1)$.

Lemma 1. There exists a positive integer k such that for every $t \in J$ and $n \in \mathbb{N}$, among the numbers $a(t,n),\ldots, a(t,n+k-1)$ there is at least one 0 and at least one 1.

Proof. Since α is irrational, the trajectory of each $t \in J$ under the iterates of R_α is dense in J. Hence, the set $\bigcap_{i=0}^{\infty} R_\alpha^{-i}([0,1/2])$ is empty. But for all $i \in \mathbb{N}$ the set $R_\alpha^{-i}([0,1/2]) \cap [0,1/2]$ is compact, and hence there exists a positive integer k such that the set $\bigcap_{i=0}^{k-1} R_{\alpha}^{-i}([0,1/2])$ is empty. This proves that for every t and every n there is at least one 1 among $a(t,n),\ldots,a(t,n+k-1)$. Since

$$a(t,n) = 1 - a(s,n) \quad \text{for} \quad s = \{t + 1/2\}$$

the same is true for 0 instead of 1. □

The following fact is simple and well-known.

Lemma 2. If $t,s \in J$ and $t \neq s$ then $a(t) \neq a(s)$. □

For each finite sequence $\underline{w} = (w_i)_{i=0}^{n-1} \in \{0,1\}^n$ and $t \in J$ we define $a_{\underline{w}}(t) \in X$ by setting $a_{\underline{w}}(t) = (y_i)_{i=0}^{\infty}$, where $y_i = w_i$ for $i = 0,1,\ldots,n-1$ and $y_i = a(\overline{t},i-n)$ for $i = n,n+1,\ldots$. For $\underline{w}$ as above and a set $A \subset \mathbb{N}$ such that $\mathbb{N}\setminus A$ is infinite, we define a map $G_{A,\underline{w}} : J \to I$ by $G_{A,\underline{w}}(t) = p(F(A,a_{\underline{w}}(t)))$.

Lemma 3. The map $G_{A,\underline{w}}$ is a Borel map.

Proof. The maps $p, F(A,\cdot)$ and the map which sends $a(t)$ to $a_{\underline{w}}(t)$, are continuous. Therefore it is enough to show that the map a is Borel. But this is obvious, since the inverse image of a cylinder is an intersection of intervals. □

For $n = 1,2,\ldots$, set $A_n = \{n^2+1, n^2+2,\ldots,n^2+n\}$ and $B_n = \bigcup_{i=1}^{\infty} A_{i\cdot 2^{n-1}}$. Clearly, for each n the set $\mathbb{N}\setminus B_n$ is infinite.

Lemma 4. For each finite 0-1 sequence $\underline{w}$ and each positive integer n, the set $G_{B_n,\underline{w}}(J)$ is a scrambled set for f.

Proof. The sequence $F(B_n,a_{\underline{w}}(t))$ arises from $a(t)$ by adding $\underline{w}$ at the beginning and then inserting blocks of 1's which are longer and longer but the distances between them are also longer and longer. Therefore, for each $t \in J$ the sequence $F(B_n,a_{\underline{w}}(t))$ has arbitrarily long blocks of 1's and (by Lemma 1) arbitrarily long blocks with no k consecutive 1's. Hence the condition (1) (for $G_{B_n,\underline{w}}(J)$ as S) is satisfied (if $p(\underline{x})$ is periodic then $\underline{x}$ is also periodic).

Let $t,s \in J$, $t \neq s$. Since for all $i \in \mathbb{N}$ we have $R_{\alpha}^{i}(t) \neq R_{\alpha}^{i}(s)$, by Lemma 2 there are infinitely many $j \in \mathbb{N}$ such that $a(t,j) \neq a(s,j)$. Hence, there are infinitely many $j \in \mathbb{N}$ such that if $\underline{y} = (y_i)_{i=0}^{\infty} = F(B_n,a_{\underline{w}}(t))$ and $\underline{z} = (z_i)_{i=0}^{\infty} = F(B_n,a_{\underline{w}}(s))$ then $y_j \neq z_j$. Since there is at least one 1 among both $y_{j+2},\ldots,y_{j+k+1}$ and $z_{j+2},\ldots,z_{j+k+1}$, we have

$$|p(h^j(\underline{y})) - 1/2| \geq 2^{-(k+2)}$$

and

$$|p(h^j(\underline{z})) - 1/2| \geq 2^{-(k+2)}$$

(if $|p(\underline{x}) - 1/2| < 2^{-(k+2)}$ then $x_i = 0$ for $i = 2,3,\dots,k+1$).

Since the 0-th coordinates of $h^j(\underline{y})$ and $h^j(\underline{z})$ are y_j and z_j respectively, one of them has to be 0 and the other one 1. Therefore, the points $p(h^j(\underline{y}))$ and $p(h^j(\underline{z}))$ lie on opposite sides of 1/2 and the distance of each of them from 1/2 is at least $2^{-(k+2)}$. Thus,

$$|p(h^j(\underline{y})) - p(h^j(\underline{z}))| \geq 2^{-(k+1)} .$$

Since $p(h^j(\underline{y})) = f^j(G_{B_n,\underline{w}}(t))$ and $p(h^j(\underline{z})) = f^j(G_{B_n,\underline{w}}(s))$, we obtain

$$\limsup_{i\to\infty} |f^i(G_{B_n,\underline{w}}(t)) - f^i(G_{B_n,\underline{w}}(s))| \geq 2^{-(k+1)} , \tag{4}$$

i.e. (2) is satisfied. This also shows that $G_{B_n,\underline{w}}$ is injective and since the set J is uncountable, the set $G_{B_n,\underline{w}}(J)$ is also uncountable.

Since there are arbitrarily long blocks of 1's on the same coordinate for $F(B_n,a_{\underline{w}}(t))$ and $F(B_n,a_{\underline{w}}(s))$, we have

$$\liminf_{i\to\infty} |f^i(G_{B_n,\underline{w}}(t)) - f^i(G_{B_n,\underline{w}}(s))| = 0 ,$$

i.e. (3) is satisfied. □

Since the set of all finite 0-1 sequences is countable, we can write a sequence $(\underline{w}(n))_{n=1}^{\infty}$ consisting of all of them. Set $F_n(t) = F(B_n,a_{\underline{w}(n)}(t))$, $G_n = G_{B_n,\underline{w}(n)}$, $S_n = G_n(J)$ and $S = \bigcup_{n=1}^{\infty} S_n$.

Lemma 5. The set S is a scrambled set for f.

Proof. The condition (1) is satisfied by Lemma 4. Take $s,t \in J$ and positive integers m,n such that $(t,n) \neq (s,m)$. If $n = m$, then conditions (2) and (3) for $x = G_n(t)$ and $y = G_m(s)$ are satisfied by Lemma 4. Assume that $n \neq m$. By the same arguments as in the proof of Lemma 4, we obtain that (3) is satisfied. Without any loss of generality we may assume that $m < n$. Then there are arbitrarily long blocks of 1's for $F_m(s)$ such that for $F_n(t)$ on the same coordinates there is 0 in every block of length k (by Lemma 1). Therefore, in the same way as in the proof of Lemma 4, we obtain

$$\limsup_{i\to\infty} |f^i(G_n(t)) - f^i(G_m(s))| \geq 2^{-(k+1)} , \tag{5}$$

i.e. (2) is satisfied. □

Denote the Lebesgue measure on the unit interval by λ. By Lemma 3, the image of λ under G_n is a Borel measure. Call this image μ_n. We have $\mu_n(I) = \mu_n(S_n) = 1$. Since G_n is injective and λ is non-atomic, μ_n is also non-atomic. Hence, the measure $\mu = \sum_{n=1}^{\infty} 2^{-n} \mu_n$ is a non-atomic Borel measure and $\mu(I) = \mu(S) = 1$.

Lemma 6. If U is an open non-empty set, then $\mu(U) > 0$.

Proof. If U is open and non-empty, so is the set $p^{-1}(U)$. Hence it contains some cylinder $C = \{(y_i)_{i=0}^{\infty} \in X : y_i = w_i \text{ for } i = 0,1,\dots,j\}$, where $\underline{w} = (w_i)_{i=0}^{j}$ is some 0-1 sequence. If m is large enough then the smallest element of B_m is larger than j. Therefore there exists n such that $\underline{w}(n)$ begins by $\underline{w}$ (and consequently, $\{(y_i)_{i=0}^{\infty} \in X: y_i = w_{i,n} \text{ for } i = 0,1,\dots,\ell\} \subset C$, where $\underline{w}(n) = (w_{i,n})_{i=0}^{\ell})$ and the smallest element of B_n is larger than j. Then we have $F_n(J) \subset p^{-1}(U)$, and consequently $S_n \subset U$. Hence, $\mu(U) \geq 2^{-n}$. □

Define a map $q : I \to I$ by $q(x) = \mu([0,x])$.

Lemma 7. The map q is a homeomorphism.

Proof. By Lemma 6, q is strictly increasing. Since μ is non-atomic, q is continuous. Clearly, $q(0) = 0$ and $q(1) = \mu(I) = 1$. Hence q is a homeomorphism. □

Lemma 8. The image of μ under q is λ.

Proof. If $0 \leq c < d \leq 1$ and ν is the image of μ under q, then $\nu([q(c), q(d)]) = \mu([c,d]) = \mu([0,d]) - \mu([0,c]) = q(d) - q(c)$. Since q maps I onto itself, we obtain $\nu = \lambda$. □

Lemma 9. Let $P,R : I \to I$ be continuous maps and let $Q : I \to I$ be a conjugacy between P and R, i.e. a homeorphism such that $R = Q \circ P \circ Q^{-1}$. Then

(a) for every $\varepsilon > 0$ there exists $\delta > 0$ such that if for some $x,y \in I$

$$\limsup_{n\to\infty} |P^n(x) - P^n(y)| \geq \varepsilon ,$$

then $\limsup_{n\to\infty} |R^n(Q(x)) - R^n(Q(y))| \geq \delta$

(b) if for some $x,y \in I$

$$\liminf_{n\to\infty} |P^n(x) - P^n(y)| = 0$$

then $\liminf_{n\to\infty} |R^n(Q(x)) - R^n(Q(y))| = 0$.

Proof. Since I is compact and Q^{-1} is continuous, Q^{-1} is uniformly continuous. Hence, for every $\varepsilon > 0$ there exists $\delta > 0$ such that if $|z-t| < \delta$ then

$$|Q^{-1}(z) - Q^{-1}(t)| < \varepsilon/2 .$$

We have $Q^{-1}(R^n(Q(x))) = P^n(x)$ and $Q^{-1}(R^n(Q(y))) = P^n(y)$ for each $x,y \in I$ and $n = 0,1,2,\ldots$, and thus (a) follows.

(b) follows from the continuity of Q and the equalities $Q(P^n(x)) = R^n(Q(x))$ and $Q(P^n(y)) = R^n(Q(y))$. □

Now we define a map $g : I \to I$ by $g = q \circ f \circ q^{-1}$. The map g is continuous as a composition of continuous maps. Since q is a conjugacy between f and g, then by Lemmas 5 and 9, the set q(S) is a scrambled set for g. Obviously, g has periodic points of all periods (they are images of periodic sequences under $g \circ p$).

By Lemma 8 and since $\mu(S) = 1$, we have $\lambda(q(S)) = 1$. This ends the proof of the Theorem.

Remark. We proved in fact that for every $x,y \in S$ with $x \neq y$,

$$\limsup_{n\to\infty} |f^n(x) - f^n(y)| \geq 2^{-(k+1)}$$

(see (4) and (5)), which is stronger than (2). Therefore in view of Lemma 9 (a), for g and q(S) we obtain also a condition stronger than that from the definition of a scrambled set, namely:

there exists $\delta > 0$ such that for every $x,y \in q(S)$ with $x \neq y$

$$\limsup_{n\to\infty} |g^n(x) - g^n(y)| \geq \delta.$$

REFERENCES

[1] Li, T.Y., Yorke, J.A.: Period three implies chaos, Amer. Math. Monthly 82, 985-992 (1975).

[2] Smítal, J.: A chaotic function with some extremal properties, Proc.Amer.Math. Soc. 87, 54-56 (1983).

[3] Smítal, J.: A chaotic function with a scrambled set of positive Lebesgue measure, Proc.Amer.Math.Soc.92 (1984) 50-54.

Michał Misiurewicz
Warsaw University, Institute of Mathematics
PKiN IX p., 00-901 Warszawa, Poland

ITERATIONS AND LOGARITHMS OF AUTOMORPHISMS OF COMPLETE LOCAL RINGS

C. Praagman+)

INTRODUCTION

In Praagman [2], I proved the following theorem, by combining results and methods of Chen [1], and Reich and Schwaiger, see Reich [4]:

Theorem 1. Let F be a $\mathbb{C}$-automorphism of $\mathbb{C}[[x_1,\dots,x_n]]$, then the following assertions are equivalent:

a) There exists a family of automorphisms $\{F_t\}_{t\in\mathbb{Q}}$, with $F_{t+s} = F_t \circ F_s$, $F_0 = I$ the identity and $F_1 = F$, such that the coefficient of x^α in $F_t(x_i)$ depends continuously on t.

b) There exists a $\mathbb{C}$-derivation of $\mathbb{C}[[x_1,\dots,x_n]]$ with values in the maximal ideal such that F = exp D.

The implication b $\Rightarrow$ a is trivial since F_t = exp tD satisfies the requirements. For the implication a $\Rightarrow$ b I used the existence of normal and smooth normal forms with respect to a particular choice of generators of the maximal ideal, and the existence of a Jordan decomposition for automorphisms and derivations. Here I present a much shorter direct proof, which does not involve a choice of generators, and therefore applies also to automorphisms of nonregular rings.

Background. Let V be an algebraic or analytic variety in $\mathbb{C}^m$, p a point on V, and ψ an isomorphism between two open neighbourhoods of p leaving p fixed. The origin of the problem treated in this paper is the following question. Does there exist a holomorphic vectorfield X in a neighbourhood of p, with X(p) = 0 such that for every regular point q close enough to p there exist holomorphic coordinates $z = (z_1,..,z_r)$ with z(0) = q, $X = \sum_i \frac{dz_i}{dt}\frac{\partial}{\partial z_i}$ and $\psi(q) = z(1)$. If such a vectorfield exists it implies of course the existence of a whole family of isomorphisms $\psi_t, t\in\mathbb{C}$ satisfying ψ_0 = identity, $\psi_1 = \psi$, $\psi_{s+t} = \psi_s \circ \psi_t$ by defining $\psi_t(q) = z(t)$. So the existence of X implies that ψ is connected to the identity in a certain sense. The inverse problem is: given such a connection, does this imply the existence of X? The first step in solving this kind of problem is trying to solve it formally.

+) The author is supported by the Netherlands Foundation of Mathematics (S.M.C.) with financial aid of the Netherlands Organization for the Advancement of Pure Research (Z.W.O.).

For this purpose it is more convenient to deal with the local ring of functions regular in p, or rather its completion. Then the above theorem gives an answer in case p is a regular point of V. The proof that I shall give below is also valid for singular points.

Definitions. Let R be a complete local noetherian ring with maximal ideal m and coefficient field $\mathbb{C}$, that is $R/m \cong \mathbb{C} \to R$. Let $R_\ell = R/m^{\ell+1}$, then $R = \varprojlim R_\ell$, with the canonical projections, which will be denoted by a subscript ℓ : $x \in R$, then $x_\ell \in R_\ell$. This notation will also be used for the projection on objects associated with R_ℓ. R_ℓ is a finite dimensional $\mathbb{C}$-linear space, and when it is equipped with the usual topology, this induces on R the topology of coefficientwise convergence. The most convenient way to handle this topology is the following criterion: $x_n \to x$ if and only if $(x_n)_\ell \to x_\ell$ for all $\ell \in \mathbb{N}$.

Aut R(resp Aut R_ℓ) will denote the Lie group of $\mathbb{C}$-automorphisms of R (resp R_ℓ) and Der (R,m)(resp Der(R_ℓ,m_ℓ) the Lie algebra of $\mathbb{C}$-derivations of R(resp R_ℓ) with values in the maximal ideal. As usual [,] will denote the Lie product: $[D_1,D_2] = D_1D_2 - D_2D_1$. The map exp: Der$(R,m) \to$ Aut R is defined by $\exp D = \sum_{k=0}^{\infty} \frac{D^k}{k!}$. It is clear that this series converges in C(R) the space of continuous $\mathbb{C}$-linear mappings from R into itself, since $\exp D_\ell = (\exp D)_\ell$ is well defined. In Praagman [2], theorem 4 it is proved that $\exp D \in$ Aut R for regular R, but the proof does not use the regularity of R.

If I denotes the identity map on R and $F \in$ Aut R, then for those ℓ for which $\|I_\ell - F_\ell\| < 1$ (for the operator norm), the map $F_\ell = - \sum_{k=1}^{\infty} \frac{(I_\ell - F_\ell)^k}{k}$ is well defined, and if $\|I_\ell - F_\ell\| < 1$ for all ℓ then $\log F = \varprojlim \log F_\ell$ is a well defined map, in fact a derivation (see again Praagman [2], theorem 4). It is a right inverse for the map exp: exp log F = F. Using the relation $\log F_1F_2 = \log F_1 + \log F_2$ if $F_1F_2 = F_2F_1$ (the counterpart of $\exp(D_1 + D_2) = (\exp D_1)(\exp D_2)$ if $[D_1,D_2] = 0$) it is possible to extend the definition of log to a wider domain, but in general log F will not be defined, even if $F \in \exp(\mathrm{Der}(R,m))$! The aim of this paper is to show that if there exists an iteration of F (more or less a path in Aut R joining F to I) then there exists a way to find a D such that exp D = F.

If G is an additive subgroup of $\mathbb{C}$, strictly containing $\mathbb{Z}$, equipped with the induced topology, then an analytic (resp. continuous) G-iteration of an automorphism F is a grouphomomorphism from G to Aut R, $t \to F_t$ such that $F_1 = F$ and such that for all ℓ the composition with the projection onto Aut R_ℓ is analytic (resp. continuous).

Continuous Rational Iterations. As usual a $\mathbb{Q}$-iteration is called a rational iteration. The generalization of Theorem 1 is:

Theorem 2. $F \in \exp(\mathrm{Der}(R,m))$ if and only if there exists a continuous rational iteration of F.

Proof. If $F = \exp D$, then $t \to \exp tD$ is an analytic rational iteration of F. So assume $t \to F_t$ is a continuous rational iteration of F, then for all $\ell \in \mathbb{N}$ there exists a t_0 such that $\| I_\ell - (F_t)_\ell \| < 1$ for $|t| < t_0$. Then $D_\ell = \frac{1}{t}\log(F_t)_\ell$ is independent of t, for let $t,s \in \mathbb{Q}$, then there exist $m,n,u \in \mathbb{Z}$ such that $s = \frac{m}{u}$, $t = \frac{n}{u}$ and $\frac{1}{s}\log(F_s)_\ell = \frac{u}{m}\log(F_{m/u})_\ell = \frac{u}{m}\log(F^m_{1/u})_\ell = u \log(F_{1/u})_\ell = \frac{u}{n}\log(F_{n/u})_\ell = \frac{1}{t}\log(F_t)_\ell$.

And $\exp D_\ell = F_\ell$: $\exp D_\ell = \exp u \log(F_{1/u})_\ell = (\exp \log(F_{1/u})^u_\ell = (F_{1/u})^u_\ell = F_\ell$. So $D = \varprojlim D_\ell$ satisfies $\exp D = F$.

The proof of Theorem 2 yields more than the stated property. So Theorem 2 may be rephrased in the following stronger statement:

Theorem 3. If $t \to F_t$ is a continuous rational iteration of $F \in \mathrm{Aut}\, R$, then there exists a $D \in \mathrm{Der}(R,m)$ such that $F_t = \exp tD$, and this D is unique.

Proof. The existence of D is proved above. Let D_1 and D_2 both satisfy the requirements. Then $\exp(tD_1 - tD_2) = F_t \circ F_t^{-1} = I$ for all $t \in \mathbb{Q}$. Differentiating this expression with respect to t in zero yields $D_1 = D_2$.

Remarks. 1. According to Reich [4] the existence of a continuous rational iteration is equivalent to the existence of a complex analytic one, but Theorem 3 implies that every continuous rational iteration may be extended to a complex analytic iteration. 2. Note that here the rationality of the iteration is not essential. As long as one has a continuous G-iteration of F, where 0 is a limitpoint of G the proof works.

Analytic Rings. In this section let R be an analytic ring over $\mathbb{C}$, that is a homomorphic image of $\mathbb{C}\{x_1,\dots,x_m\}$ for some m. Assume there exists a continuous group homomorphism $\mathbb{Q} \to \mathrm{Aut}\, R$, $t \to F_t$, $F_1 = F$. As above this will be called a continuous rational iteration of F. Then the composition $\mathbb{Q} \to \mathrm{Aut}\, \hat{R}$, $\hat{R}$ the completion of R with respect to the m-adic topology, is a continuous rational iteration of F in $\mathrm{Aut}\, \hat{R}$. Hence $F_t = \exp tD$. Since $D = \frac{\partial F}{\partial t}\big|_{t=0}$, $D \in \mathrm{Der}(R,m)$. So we have

Theorem 4. Let R be an analytic ring over $\mathbb{C}$, $F \in \mathrm{Aut}\, R$, and $t \to F_t$ a continuous rational iteration of F. Then there exists precisely one derivation $D \in \mathrm{Der}(R,m)$ such that $F_t = \exp tD$.

Note that the same kind of things could be said about homomorphic images of $\mathbb{C}[x_1,\ldots,x_m]_{(x_1,\ldots,x_m)}$, the localization to $(x_1,\ldots,x_m)$, but it is not hard to see that exp D = F can only occur if D and F are linear: $Dx_i = \Sigma\alpha_j x_j$, $Fx_i = \Sigma\beta_j x_j$, in which case the possibilities are well known.

Weaker Conditions. In a forthcoming paper, Praagman [3], I shall prove that the condition that the rational iteration is continuous is superfluous, and that even the group structure is superfluous (R is complete again):

Theorem 5. Let $F \in$ Aut R, assume that for all $n \in \mathbb{N}$ the equation $G^n = F$ has a solution $G \in$ Aut R. Then $F \in \exp(\text{Der } R, m)$.

In that paper I shall also investigate the set $\exp(\text{Der}(R, m))$ further.

REFERENCES

[1] Chen, K.T.: Local diffeomorphisms - C^∞-realization of formal properties. Am. J. Math. 87, 140-157 (1965).

[2] Praagman, C.: Iterations and logarithms of formal automorphisms. Aeq. Math. 28 (1985).

[3] Praagman, C.: Automorphisms of complete local rings having a logarithm. Preprint Groningen (1985).

[4] Reich, L.: Iteration problems in Power Series Rings. Théorie de l'iteration et ses applications. Coll. Int. du CNRS 332, Paris 1982.

C. Praagman
Onderafdeling der Wiskunde en Informatica
Technische Hogeschool Eindhoven
Postbus 513, 5600 MB Eindhoven
Nederland

ON A DIFFERENTIAL EQUATION ARISING IN ITERATION THEORY IN RINGS OF FORMAL POWER SERIES IN ONE VARIABLE

L. Reich

1. INTRODUCTION OF THE DIFFERENTIAL EQUATION

Let $\mathbb{C}[[x_1,\dots,x_n]]$ be the ring of formal power series with complex coefficients in n indeterminants $x_1,\dots,x_n$. In this paper we are going to consider automorphisms of $\mathbb{C}[[x_1,\dots,x_n]]$ which are continuous in the order topology of $\mathbb{C}[[x_1,\dots,x_n]]$ and leave every element of the field $\mathbb{C}$ fixed. These automorphisms are uniquely determined by the image of $x = {}^T(x_1,\dots,x_n)$, $F(x) = Ax + p(x)$ where A is a non-singular (n,n)-matrix and p(x) is the so called non-linear part of F. Moreover, these automorphisms form a group Γ under composition o, which is, in the above representation F(x), the same as substitution of the invertible vector F(x) of formal power series. As references for these concepts and the notions of analytic, continuous and fractional iteration in Γ, as well as the basic results referring to this we mention [1,2] or [3], pp.155-166. Now let $(F_t)_{t \in \mathbb{C}}$ be an analytic iteration (group) of a given $F \in \Gamma$. By writing more explicitely $F_t(x) = F(t,x) = \sum_{\nu, |\nu| > 1} p_\nu(t)x^\nu$ (with $x^\nu = x_1^{\nu_1}\dots x_n^{\nu_n}$ for $\nu = (\nu_1 \dots \nu_n)$, we have, by definition of an iteration:

$$F(t,F(s,x)) = F(t+s,x)(= F(s,F(t,x)) , \tag{1}$$

$$F(1,x) = F_1(x) = F(x) , \tag{2}$$

$$\text{all functions } p_\nu : t \to p_\nu(t) \in \mathbb{C}^n \tag{3}$$

are entire functions. Hence, differentiation with respect to t is possible and the chain rule for formal power series yields

$$\frac{\partial F}{\partial s}(t+s,x) = \frac{\partial F}{\partial y}(t,y)\bigg|_{y=F(s,x)} \cdot \frac{\partial F(s,x)}{\partial s} \tag{3}$$

where $\frac{\partial F}{\partial y}(t,y)$ denotes the Jacobian matrix $\left(\frac{\partial F_j(s,y)}{\partial y_k}\right)_{1<j<k}$ and $\frac{\partial F}{\partial s}(s,x)$ is performed coefficient-wise. Putting s = 0, we obtain, observing that in an iteration group necessarily F(0,x) = x:

$$\frac{\partial F}{\partial t}(t,x) = \frac{\partial F(t,x)}{\partial x} \cdot G(x) ,$$

where $G(y) = \frac{\partial F(0,y)}{\partial s}$ is usually called the infinitesimal generator of $(F_t)_{t \in \mathbb{C}}$, more precisely, the function determining the infinitesimal generator of the semigroup of operators defined by $\Omega^t\psi = \psi \circ F_t$. The left hand side of the above relation may still be simplified further. We recall that according to (1):

$$\frac{\partial}{\partial s}F(t+s,x) = \frac{\partial}{\partial s}F(s,F(t,x)) .$$

Putting again s = 0 and using the definition of G once more we see that eventually we end up with the (in general partial) functional-differential equation:

$$G(F_t(x)) = \frac{\partial F_t(x)}{\partial x} \cdot G(x) \quad \text{(for all } t \in \mathbb{C}) ; \tag{4}$$

and hence in particular for t = 1:

$$G(F(x)) = \frac{\partial F}{\partial x} \cdot G(x) . \tag{5}$$

This differential equation is well known in iteration theory. It seems to occur in this connection for the first time in E.Jabotinsky's papers [4] and [5]; see also Gy.Targonski's Graz lectures [6], p.34. (5) is even sometimes used to solve the iteration (i.e. embedding) problem, e.g. in [7]. But it is not known whether the theory of analytic iterations in rings of formal power series as initiated in [8,9, 10], and developed in [11,12] and briefly sketched in [1] and [2] may entirely be rebuilt on the base of (5). But nevertheless, equation (5) seems to be of interest for its own sake. We will therefore, for n = 1, i.e. for power series in one variable, give the complete solution of (5) and describe the group structure of the general solution in some detail. This is the contents of Theorem 2 (for Jabotinsky's differential equation in its so called normal form) and of Theorem 3 in the general situation.

But before doing so we have to explain what equation (5) means for the classification of formal autonomous differential systems. Let us consider, this purpose in mind, a formal autonomous differential system

$$\frac{dy}{dt} = G(y) , \tag{6}$$

where $G(y) \in (\mathbb{C}[[y_1,\dots,y_n]])^n$ is given. These objects were introduced by G.D.Birkhoff ([13]), see also [11]. Suppose now that $T \in \Gamma$, $y = T(z)$, transforms (6) into itself. The chain rule for formal power series gives us

$$\frac{\partial T}{\partial z} \cdot \frac{dz}{dt} = G(T(z)) ,$$

and since we require that also

$$\frac{dz}{dt} = G(z) ,$$

we find for T, as a necessary and sufficient condition the equation (5) for T(z):

$$G(T(z)) = \frac{\partial T}{\partial z} \cdot G(z) . \tag{5}$$

It is obvious that the solutions T form a subgroup of Γ, i.e. a group under substitution, since clearly all $T \in \Gamma$ which leave (6) invariant form a group. This may also be shown directly, as follows. If

$$(G \circ F_1)(x) = \frac{\partial F}{\partial x} \cdot G(x)$$

and

$$(G \circ F_2(x) = \frac{\partial F_2}{\partial x} \cdot G(x) ,$$

then

$$(G \circ F_1 \circ F_2)(x) = \left.\frac{\partial F_1}{\partial x}\right|_{y = F_2(x)} \cdot G(F_2(x)) =$$

$$= \left.\frac{\partial F_1}{\partial x}\right|_{y = F_2(x)} \cdot \frac{\partial F_2}{\partial x} \cdot G(x) = \frac{\partial (F_1 \circ F_2)}{\partial x} \cdot G(x) . \tag{7}$$

Moreover, if F fulfills (5), then, by differentiating $(F^{-1} \cdot F)(x) = E(x)$, we obtain

$$\left.\frac{\partial F^{-1}}{\partial x}\right|_{y = F(x)} \cdot \frac{\partial F}{\partial x} = E ,$$

and therefore

$$G(x) = G(F \circ F^{-1})(x) = (G \circ F)(F^{-1}(x)) =$$

$$= \left.\frac{\partial F}{\partial y}\right|_{y = F^{-1}(x)} \cdot G(F^{-1}(x)) ,$$ which is clearly (5) for F^{-1}.

2. THE GENERAL SOLUTION OF JABOTINSKY'S DIFFERENTIAL EQUATION IN THE CASE OF ONE VARIABLE

Let us begin with the right hand side $G(x) = \lambda x + c_2x^2 + \dots$, where $\lambda \neq 0$. If we are interested in the group structure of the general solution, we may replace the equation

$$\frac{dy}{dt} = G(y) \tag{6}$$

by any equivalent system which is related to (6) by a transformation $y = T(z)$, $T \in \Gamma$. The same calculations as in 1. yield the relation

$$\frac{dT}{dz} \cdot G(T(z)) = H(z) \quad .$$

It is well known (cf.[11]) that we may choose the linear normal form

$$H(z) = \lambda z \ .$$

Therefore, instead of (5) we are looking now for the solutions $F(z)$ of

$$\lambda F(z) = \lambda z \cdot \frac{\partial F}{\partial z} \ .$$

Inserting $F(z) = c_1z + c_2z^2 + \dots$ we find

$$\lambda c_j = j.\lambda.c_j \ , \quad j = 1,2,\dots$$

which means that $c_1 \neq 0$ may be chosen arbitrarily, whereas $c_j = 0$, for $j \geq 2$. Hence we may summarize:

<u>Theorem 1:</u> If in (5) we have $G(z) = \lambda z + a_2z^2 + \dots$, where $\lambda \neq 0$, then the solutions $F(z) = \rho z + d_2z^2 + \dots$ $(\rho \neq 0)$ form an abelian group under composition. The group is isomorphic to $\mathbb{C}^\cdot$, the multiplicative group of $\mathbb{C}$, via $F(z) = \rho z + \dots \to \rho$. To each $\rho \in \mathbb{C}^\cdot$ there corresponds a unique solution $F(z) = \rho z + d_2z^2 + \dots$, the coefficients being universal polynomials p_k in $\rho, a_2, \dots a_k$. If $G(z)$ is taken in its normal form λz, then all solutions of (5) are linear functions.

The case $G(z) = c_mz^m + \dots$, $m > 1$, $c_m \neq 0$, is much more interesting. Firstly, we need a Lemma on the normal form of G in this case.

<u>Lemma 1.</u> If $G(y) = c_my^m + \dots$, $m > 1$, $c_m \neq 0$, then equation (5) is equivalent to

$$\frac{dz}{dt} = z^m + dz^{2m-1}, \quad d \in \mathbb{C} \ .$$

Proof. Take $y = t_1 z$. Then,

$$\frac{dz}{dt} = t_1^{m-1} c_m z^m + \dots \quad .$$

Choosing t_1 such that $t_1^{m-1} = c_m^{-1}$, we have

$$\frac{dz}{dt} = z^m + c'_{m+1} z^{m+1} + \dots \quad .$$

Therefore, we may assume that already $c_m = 1$. Now let $y = T(z)$ be $z + \delta_2 z^2 + \dots$, then (7) leads us to

$$\frac{dT}{dz}(z^m + dz^{2m-1}) = (z+\delta_2 z^2 + \dots)^m + c_{m+1}(T(z))^{m+1} + \dots$$

or

$$(1 + 2\delta_2 z + 2\delta_3 z^2 + \dots)(z^m + dz^{2m-1}) =$$

$$= (z + \delta_2 z^2 + \dots)^m + \sum_{k=1}^{\infty} c_{m+k}(T(z))^{m+1} \quad .$$

The smallest possible value of m is $m = 2$. Then $2m-1 = m+1 = 3$. Comparing the coefficients of z^ν on both sides of the last equation we see that equality already holds for $\nu = 2$. For $\nu = 3$ we find

$$2\delta_2 + d = 2\delta_2 + c_3 \quad .$$

Here δ_2 remains undetermined whereas $c_3 = d$.

For $k \geq 3$ we obtain

$$(k-2)\delta_k = R_k(d,\delta_2,\dots,\delta_{k-1}) \ ,$$

and hence by induction

$$\delta_k = P_k(d,\delta_2), (k \geq 3) \ ,$$

where P_k is a polynomial. Suppose now $m > 2$. Then by similar calculations we see that equality holds for the coefficients of z^m. For $z^{m+1},\dots,z^{2m-2}$ we find

$$(k-m)\delta_k = R_k(\delta_2,\dots,\delta_{k-1},c_j), \quad k = 2,\dots,m-1 \ ,$$

and hence these δ_k are uniquely determined. On the other hand, if $k = m$, we obtain

$$m\delta_m + d = m\delta_m + c_{2m-1} ,$$

so that δ_m may be chosen arbitrarily, and $d = c_{2m-1}$. For $k > m$ we have again the formula

$$(k-m)\delta_k = R_k(d,\delta_2,\ldots,\delta_m,\ldots,\delta_{k-1})$$

where R_k is a polynomial, and therefore these δ_k's are uniquely determined once δ_m is chosen. Therefore transformations to the normal form exist.

We will now prove the following result, basic for our main theorem.

Theorem 2. Suppose that the differential equation for $F(z) = a_1z + a_2z^2 + \ldots$, $a_1 \neq 0$,

$$F(z)^m + d(F(z))^{2m-1} = (z^m + dz^{2m-1})\frac{dF}{dz} \tag{5'}$$

with $m > 1$ is given. Then the following assertions are true:

(i) $a_1^{m-1} = 1$.

(ii) To each solution ρ of $a_1^{m-1} = 1$ there exist solutions of (5'),

$$F(z) = \rho z + a_2z^2 + \ldots ,$$

and we have necessarily

$$F(z) = \rho z + \sum_{\nu=1}^{\infty} P_\nu(\rho,a_m)z^{\nu(m-1)+1} , \tag{8}$$

where the P_ν's are universal polynomials in a_m, and $P_1 \equiv a_m$. The coefficient a_m may be chosen arbitrarily, and to ρ and a_m there exists exactly one solution (8), denoted by $F(d;\rho;a_m;z)$.

Proof. $F(z) = a_1z + a_2z^2 + \ldots$, with $a_1 \neq 0$, and equation (5) for $F(z)$ yield for a_1 the "characteristic equation"

$$a_1^{m-1} = 1 ,$$

hence $a_1 = \exp(2\pi i \frac{k}{m-1})$, $0 \leq k < m-1$, where, as we will show, each determination of the (m-1)-th root of unity a_1 leads to solutions of (5'). Let us now suppose $m = 2$, i.e. $m-1 = 1$, and hence $a_1 = 1$. Then (5') is the same as

$$(z + a_2z^2 + a_3z^3 + \dots)^2 + d(z + az^2 + a_3z^3 + \dots)^3 =$$

$$= (1 + 2a_2z + 3a_3z^2 + \dots)(z^2 + dz^3)\,.$$

Equating coefficients of z^3 on both sides leads to the relation

$$2a_2 + d = d + 2a_2\ ,$$

which is true for all values of a_2. Choose a fixed, but arbitrary $a_2 \in \mathbb{C}$. Then proceeding to coefficients of higher order we find

$$(2-j)a_j = p_j(d,a_1,a_2,\dots,a_{j-1}),\quad j \geq 3\ ,$$

where the p_j's are universal polynomials. Hence, by induction, we get, for $j \geq 3$,

$$a_j = P_j(a_1,a_2,d)\ ,$$

where the P_j are again universal polynomials.

Suppose now $m > 2$, hence $m-1 > 1$. Here we have, in each case, more than one possibility for a_1 at our disposal. Let us fix one possible choice of a_1. We have to investigate

$$(a_1z + a_2z^2 + \dots)^m + d(a_1z + a_2z^2 + \dots)^{2m-1} =$$

$$= (1 + 2a_2z + \dots)(z^m + dz^{2m-1})\ . \tag{9}$$

We shall prove that for $j = 2,\dots,m-1$,

$$a_j = 0$$

holds. For a_2 we obtain indeed

$$ma_1^{m-1}\,a_2 = 2a_2\ ,$$

or, since $a_1^{m-1} = 1$, $2a_2 = ma_2$, and consequently, $a_2 = 0$. If for $j = 2,\dots,k-1$, and $k \leq m-1$, $a_j = 0$ is already proved, then, by comparing the coefficients of z^{m+k}, on both sides of (9), we see that

$$(m-k)a_k = 0\ ,$$

or $a_k = 0$. So our first assertion is established.

Now, consider the equation for a_m. By looking at z^{2m-1} and using once more $a_1^{m-1} = 1$, and $a_2 = \dots = a_{m-1} = 0$,

$$m a_1^{m-1} a_m + d a_1^{2m-2} a_1 = m a_m + d a_1$$

yields that a_m is arbitrary. Taking any value of a_m and applying $a_2 = 0$, we deduce from

$$m . a_{m+1} a_1^{m-1} = (m+1) a_{m+1} + 2a_2 . d$$

that $a_{m+1} = 0$, and, similarly by induction, $a_j = 0$, for $j = m+1,\dots,2m-2$, whereas

$$a_{2m-1} = P_{2m-1}(d;a_1;a_m)$$

where P_{2m-1} is a universal polynomial. Now, proceeding by induction over ν we prove by quite analogous calculations that

$$a_{1+\nu(m-1)} = P_{1+\nu(m-1)}(d,a_1,a_m) \ ,$$

$$a_{1+\nu(m-1)+j} = 0, \quad \text{for} \quad 1 \le j \le m-2 \ .$$

Therefore, Theorem 2 is proved. Let us note that the form

$$a_1 x + a_m x^m + \sum_{\nu=2}^{\infty} P_{1+\nu(m-1)}(d,a_1,a_m) x^{1+\nu(m-1)}$$

of the solutions is nothing else than the so called semicanonical form of a formal power series in one variable, of the type

$$e^{2\pi i \frac{1}{m-1}} x \quad + \dots \quad . \qquad \text{(cf. [11])}$$

Now we know all about the solutions of (5') in order to prove a result on the group structure of the set of all solutions.

Lemma 2. Let us denote by $\Gamma(m,d)$ the group of solutions of (5'), the group operation being substitution. Then $\Gamma(m,d)$ is an abelian group, isomorphic to the multiplicative group of all matrices

$$\begin{pmatrix} a_1 & 0 \\ a_m & a_1 \end{pmatrix} , \quad a_1^{m-1} = 1 \ , \quad a_m \in \mathbb{C} \ \text{arbitrary} \quad .$$

The isomorphism is given by the mapping

$$F(d,a_1,a_m,x) \to \begin{pmatrix} a_1 & 0 \\ a_m & a_1 \end{pmatrix} .$$

Proof. Let

$$F(x) = a_1x + a_mx^m + \sum_{\nu\geq 2} p_{1+(m-1)}(d,a_1,a_m)x^{1+\nu(m-1)}$$

and

$$F^*(x) = a_1^*x + a_m^*x^m + \sum_{\nu\geq 2} p_{1+\nu(m-1)}(d,a_1^*,a_m^*)x^{1+\nu(m-1)}$$

be two solutions of (5'). Then we know already that $F \circ F^*(x)$ and $F^* \circ F(x)$ are solutions, too, and both have the m-jet

$$a_1a_1^*x + (a_1^*a_m + a_1a_m^*)x^m \ \ldots \qquad .$$

But since a solution of (5') is uniquely determined by the coefficients of x and of x^m, we have $F^* \circ F(x) = F \circ F^*(x)$ which means that $\Gamma(m,d)$ is abelian.

Lemma 3. If we denote by $\Gamma^1(m,d)$ the subgroup of $\Gamma(m,d)$ consisting of all solutions of (5') of the form

$$z + a_mz^m + \ldots$$

then the quotient group $\Gamma(m,d)/\Gamma^1(m,d)$ is cyclic of order m-1 and each coset may be described as a subset of $\Gamma(m,d)$ consisting of all solutions of the form

$$\exp\left(\frac{2\pi i}{m-1}\cdot k\right)x + a_mx^m + \ldots$$

with fixed k ($0 \leq k \leq m-1$).

Proof. Clearly, $\Gamma^1(m,d)$ is a subgroup. If F,G belong to the same coset modulo $\Gamma^1(m,d)$ then

$$F^{-1} \circ G(x) = x + b_mx^m + \ldots ,$$

so that

$$F(x) = a_1x + a_mx^m + \ldots$$

and

$$G(x) = a_1 x + b_m x^m + \dots ,$$

and vice versa. Since there are m-1 solutions of $a_1^{m-1} = 1$, the group $\Gamma(m,d)/\Gamma^1(m,d)$ is of order m-1. Moreover, if

$$F(x) = e^{2\pi i \frac{k}{m-1}} x + a_m x^m + \dots$$

and

$$G(x) = e^{2\pi i \frac{1}{m-1}} x + b_m x^m + \dots$$

then clearly F(x) and $G^k(x)$ belong to the same coset, so that the group Γ/Γ^1 is cyclic. Therefore Lemma 2 is proved.

So far we have only considered Jabotinsky's equation in its normal form. But the results of Theorem 2, Lemma 2 and Lemma 3 are essentially invariant under conjugation of solutions and equivalence of differential systems.

Now suppose that $T: x \to x + t_z x^2 + \dots$ transforms the differential equation

$$(G \circ F)(x) = G(x) \cdot \frac{\partial F}{\partial x} \tag{5}$$

with

$$G(x) = x^m + c_{m+1} x^{m+1} + \dots$$

to its normal form

$$(\tilde{G} \circ \tilde{F})(x) = \tilde{G}(x) \cdot \frac{\partial \tilde{F}}{\partial x} , \tag{10}$$

where $\tilde{G}(x) = x^m + dx^{2m-1}$. We have now to show in detail which kind of action this transformation of x induces on the integrals of (5). As we already said, it induces a 1-1-relation by conjugation between the solutions of (5) and (10). This is completely described by

<u>Lemma 4.</u> If $T: x \to x + t_2 x^2 + \dots$ transforms (5) to its normal form (10) then

$$F \to \tilde{F} = T^{-1} \circ F \circ T$$

is a bijective mapping of the set of integrals of (5) onto the set of integrals of (10).

<u>Proof.</u> From (5) we deduce

$$(G \circ F)T(x)) = \frac{\partial F}{\partial x}(T(x)) \cdot G(T(x))$$

and

$$(G \circ T)(T^{-1} \circ F \circ T)(x) = \frac{\partial F}{\partial x}(T(x)) \cdot (G \circ T)(x) .$$

But since

$$\frac{\partial (T^{-1} \circ F \circ T)(x)}{\partial x} = \frac{\partial (T^{-1})}{\partial u}\bigg|_{u=F \circ T(x)} \cdot \frac{\partial F}{\partial z}\bigg|_{z=T(x)} \cdot \frac{\partial T}{\partial x}$$

and

$$\frac{\partial (T^{-1})}{\partial u}\bigg|_{u=F \circ T(x)} = \left(\frac{\partial T}{\partial v}\right)^{-1}\bigg|_{v=T^{-1} \circ F \circ T(x)} ,$$

we eventually get, after multiplying the differential equation by $\frac{\partial T^{-1}}{\partial u}\Big|_{u=F \circ T(x)}$:

$$\left| \left(\frac{\partial T}{\partial v}\right)^{-1} \cdot (G \circ T)(v) \right|_{v=T^{-1} \circ F \circ T(x)} = \frac{\partial (T^{-1} \circ F \circ T)}{\partial x} \cdot \left(\frac{\partial T}{\partial x}\right)^{-1} (G \circ T)(x) .$$

If we introduce the abbreviation

$$\tilde{F} = T^{-1} \circ F \circ T$$

and take into account that (cf. the proof of Lemma 1)

$$\tilde{G}(x) = \left(\frac{\partial T}{\partial x}\right)^{-1} (G \circ T(x) ,$$

then we obtain the Lemma.

We have already shown that

$$\tilde{F}(x) = \rho x + \tilde{\gamma}_m x + \sum_{\nu=2}^{\infty} Q_\nu(\rho, \tilde{\gamma}_m) x^{1+\nu(m-1)} ,$$

where $\tilde{\gamma}_m$ can be any complex number and $\rho^{m-1} = 1$. A lengthy but elementary calculation gives us the information that, with $F(x) = (T \circ \tilde{F} \circ T^{-1})(x)$,

$$F(x) = \rho x + \sum_{\mu=2}^{m-1} p_\mu(\rho, t_2, \ldots, t_\mu) x^\mu +$$

$$+ (\tilde{\gamma}_m + q_m(\rho, t_2, \ldots, t_m)) x^m + \sum_{\nu \geq m+1} p_\nu(\rho, \tilde{\gamma}_m, t_k) x$$

where $p_2,\ldots,p_{m-1},q_m$ are polynomials independent of $\tilde{\gamma}_m$. Hence, $\gamma_m = \tilde{\gamma}_m + q_m$ can also be taken as an arbitrary complex number, and $\tilde{\gamma}_m$ is uniquely determined by γ_m and vice versa. The coefficients $p_\nu(\tilde{\gamma},\gamma_m,t_k)$, for $\nu > m$, can be calculated recursively. These remarks may be proved as follows. Let $\tilde{F}(x)$ be a solution of (10), of the type $\tilde{F}(x) = x + \tilde{d}x^m + \ldots$ Since $T^{-1} \circ T(x) = x$, the (m-1)-jets of $\tilde{F}(x)$ and $T^{-1} \circ \tilde{F} \circ T(x) = F(x)$ coincide because they are the same as the (m-1)-jet of $T^{-1}T(x) = x$. Consequently $F(x) = x + dx^m + \ldots$, and F is a solution of (5). If F_1,F_2 are solutions of (5) with the same linear part ρx, then $F^{-1} \circ F_2(x)$ is of the form $x + dx^m + \ldots$, and a solution, therefore the (m-1)-jets of F_1 and of F_2 are the same, and the coefficients of $x^2,\ldots,x^{m-1}$ are independent of the coefficient γ_m of x^m. But the coefficients of x^ν, $\nu > m$, are obviously polynomials in γ_m, and uniquely determined once γ_m is chosen.

Now we are able to present the main result of our paper, simply by summarizing our previous results.

Theorem 3. Consider Jabotinsky's differential equation for $F(x) = \rho x + \gamma_2 x^2 + \ldots$, $\rho \neq 0$,

$$(G \circ F)(x) = \frac{\partial F}{\partial x} \cdot G(x) \tag{5}$$

where

$$G(x) = x^m + c_{m+1}x^{m+1} + \ldots \; (m \geq 2) .$$

Then the following is true:

(i) We have necessarily $\rho^{m-1} = 1$. For each determination of ρ there exist solutions $F(x) = \rho x + \gamma_2 x^2 + \ldots$.

(ii) The coefficients γ_j, for $j = 2,\ldots,m-2$, are universal polynomials in ρ and in certain coefficients $c_k (k \leq j)$ of G.

(iii) If

$$F(x) = \rho x + \sum_{\mu=2}^{m-1} \gamma_j x^j + \gamma_m x^m + \gamma_{m+1} x^{m+1} + \ldots$$

then γ_m may be chosen arbitrarily, and the coefficients γ_ν of order ν greater than m are universal polynomials in γ_m,ρ, and $c_{m+1},\ldots,c_\nu$. Therefore the solution is uniquely determined by ρ and γ_m.

(iv) A solution of the type $F(x) = x + \ldots$ is already of the form $F(x)=x+\gamma_m x^m+\ldots$.

(v) The set of all solutions of (5) forms an abelian group $\Gamma(G)$ under substituion. If $\Gamma_1(G)$ denotes the subgroup of all solutions $F(x)=x+\gamma_m x^m+\ldots$, then the group $\Gamma(G)/\Gamma_1(G)$ is cyclic of order m-1. Each coset of $\Gamma(G)/\Gamma_1(G)$ may be described as the set of all solutions of (5) of the type $F(x) = \rho x + \ldots$, with a fixed $\rho = \exp(\frac{2\pi i}{m-1} k)$, $0 \leq k < m-1$.

(vi) The group $\Gamma(G)$ is isomorphic to the group of matrices defined in Lemma 2.

(vii) If (5) is transformed to its normal form (10) by a transformation $x \to T(x)$, then all solutions are simultaneously transformed to the type

$$\tilde{F}(x) = \rho x + \tilde{\gamma}_m x^m + \sum_{\nu>1} p_\nu(\rho,\tilde{\gamma}_m) x^{1+\nu(m-1)}$$

which is the "semicanonical form" ([11]) of the solutions $H(x)=e^{2\pi i/m-1}x + \ldots$ of (5).

REFERENCES

[1] Reich, L.: Analytische und fraktionelle Iteration formal-biholomorpher Abbildungen. Jahrbuch Überblicke der Mathematik, 123-144 (1979). Bibliographisches Institut 1979.

[2] Reich, L.: Iteration problems in power series rings. In: Théorie de l'itération et ses applications, Toulouse, 17-22, Mai 1982. Colloques internationaux du CNRS, Paris 1982.

[3] Targonski, Gy.: Topics in Iteration Theory. Vandenhoeck & Ruprecht, Göttingen 1981.

[4] Jabotinsky, E.: Iteration. Doctoral Dissertation. The Hebrew University, Jerusalem, 1-120 (1955).

[5] Jabotinsky, E.: Analytic iteration. Trans. AMS 108, 457-477 (1963).

[6] Targonski, Gy.: New directions and open problems in iteration theory. Berichte der mathematisch-statistischen Sektion im Forschungszentrum Graz, Ber.Nr.229 (1984).

[7] Beyer, W., and Channell, P.J.: A functional equation for the embedding of a homeomorphism of the interval into a flow. Lecture at the International Symposium on Iteration Theory and its Functional Equations, Lochau, Austria, 27.9.-2.10.1984.

[8] Lewis, D.N.: On Formal Power Series Transformations. Duke Journal Math. 5, 794-805 (1939).

[9] Sternberg, S.: Infinite Lie Groups and Formal Aspects of Dynamics. Journ. Math. Mech. 10, 451-474 (1961).

[10] Chen, K.T.: Local Diffeomorphisms-$\mathbb{C}^\infty$-Realizations of Formal Properties. American Journ. Math. 87, 140-157 (1965).

[11] Reich, L., and Schwaiger, J.: Über einen Satz von S. Sternberg in der Theorie der analytischen Iterationen. Monatsh. Math. 83, 207-221 (1977).

[12] Reich, L., and Schwaiger, J.: Eine Linearisierungsmethode für Funktionalgleichungen vom iterativen Typus in Potenzreihenringen. Aequationes Math. 20, 224-243 (1980).

[13] Birkhoff, G.D.: Surface Transformations and their Dynamical Applications. Acta Math. 43, 1-119.

L. Reich
Institut für Mathematik
Universität Graz
Brandhofgasse 18
A - 8010 Graz
Austria

LONG LINE ATTRACTORS

O.E. Rössler

1. INTRODUCTION

Let me offer a number of heuristic considerations, drawing on philosophy and physics beside mathematics.

The two most interesting possibilities encountered in iteration theory are, (1) the existence of continuous iterates in some maps, and (2) the existence of chaos in some others; cf.[1]. It therefore appeared of interest to see whether a connection between these two types of result cannot be established. There is one aspect to the notion of continuous fractional iterates which may be useful also in the theory of chaos-producing maps: Real numbers make sense not only when used as fractional iteration indices less than unity (cf.[2]), but also when used as the inverses of the former (that is, as ordinary iteration indices). Here, the special case of infinite reals is of particular interest. Its usefulness is not restricted to maps possessing fractional iterates.

At first sight, this idea of postulating a "transfinite accuracy" (with non-finitistic iteration indices) for iteration theory, in general, appears like an unnecessary complication. In the following, therefore, first an example from chaos theory will be presented which does call for such an extension. Thereafter, a prototypic invertible, area-contracting map will be introduced, and some of the properties of its attractor considered.

2. TRANSFINITE PERIODICITY IN ONE-DIMENSIONAL MAPS

An example is the familiar logistic recurrence, $x_{n+1} = a(x_n - x_n^2)$. For this type of maps, Sharkovsky [3] first showed that the one-dimensional parameter space (spanned by a) contains infinitely many period-doubling bifurcation sequences, all of them (along with their accumulation points) packed into a finite interval (3.57 to 4). The first sequence (which alone will be considered here) involves all periodicities that are powers of 2. In other words, it is the sequence 2^n through all n. Successive bifurcations in this sequence obey a scaling law [4]: the closer one gets to the accumulation point a_s, the closer the ratio between two successive bifurcation points approximates a certain real number (Feigenbaum's constant 4.669...). The limit point a_s thus behaves like a linear fixed point.

The reason for this scaling behaviour is that the closer one gets to the fixed point in parameter space, the more similar to each other become successive higher iterates of the map, at the given parameter value. Full identity (except for rescaling) is reached only at the accumulation point a_s proper. To the left of it, a higher iterate always has slightly less "overlap" than its immediate predecessor; to the right of it, it always has slightly more overlap ("principle of amplification of overlap" [5]). Since a minimum overlap is necessary for chaos, chaos at first appears at the highest iterate (just to the right of a_s) in order then to descend to the lower ones.

Therefore, a second bifurcation tree should be present to the right of the former. One would expect it to look a bit like a mirror image to the former (at least in the neighbourhood of a_s) - if indeed a single parameter (difference in overlap), passing monotonically across the accumulation point from negative to positive values, is responsible for all bifurcations occurring.

This "second tree" indeed exists [6,7], obeying the same scaling law as the first (see [8,9]). The difference is only that the bifurcations do not pertain to the underlying periodicity of an attracting fixed point this time, but rather to the underlying periodicity of an attracting periodic regime. An overlap just sufficient to generate chaos climbs <u>down</u> from the highest iterate where it first appeared (just to the right of a_s as mentioned) in period-halving steps in precisely the same manner as an overlap able to support a period-2 fixed point climbed <u>up</u> on the left. (Eventually when the last, the second, iterate is reached, a first periodic solution divisible by an odd integer appears [3,10] - which empirically means that "syncopes" appear in the dynamics [5]).

All of this was mentioned only to draw attention to one point: the strictly monotone passage across a_s of the "deviation from self-similarity" from negative to positive values. Strict self-similarity never applies except for a single point, a_s proper. This point in parameter space therefore plays a singular ("organizing", cf. [11]) role. Deleting it is not advisable.

A more positive statement can also be made - albeit only as an heuristic consideration. It goes like this... <u>Not</u> deleting a_s from the points of interest means, however, that the pertinent dynamical behaviour (shown by the map at this particular parameter value) also must be taken seriously. At the accumulation point proper, the periodicity of the attracting periodic solution, 2^n, goes through a singular value of n. This singular value is not finite (since otherwise a "short cut," from a finite-periodicity point attractor to a finite-underlying-periodicity chaotic attractor, would exist). Both to the left and to the right of a_s, the "same" sequence of n bifurcation points, having a_s as its limit point, applies along the a-axis. Just as in the symmetrical double sequence $\pm(1/2)^n$, with n all positive integers, the point 0 is never reached for finite n from either side, so never is a_s. <u>Adding</u> the limit point to the two open sequences that are separated by it, however, in either case amounts to formally admitting an infinite n...

If n is equal to infinity (that is, $\aleph_0$), however, 2^n is equal to $\aleph_1$. Thus, there exists a periodic attractor of an "uncountable periodicity" at a_s. (Note that this result depends on the above heuristic consideration).

That is, there exist chaos-generating maps which for certain values in their parameter spaces posses periodic attractors that can be resolved into an attracting fixed point (which they are by definition), only if one is ready to go to an iteration index that is not finite but transfinite.

Now it is known that nonperiodic solutions (constituting the bulk of solutions in a chaotic regime [10]) certainly have a "periodicity" that is larger than that of any periodic solution. This may be turned into an argument that use of transfinite iteration indices is justified, not only for a zero-measure set of points in the parameter space of the logistic recurrence as shown, but whenever there exist chaotic trajectories in arbitrary maps.

3. AN IDEALIZED AXIOM-A MAP

Realistic systems are governed by maps that are everywhere (or at least almost everywhere) invertible. The first invertible chaos-generating map ("mixing transformation") was indicated by E. Hopf [12], the so-called baker's transformation:

$$x_{n+1} = 2x_n - \Theta(x_n - 1/2)$$

$$y_{n+1} = (1/2)[y_n + \Theta(x_n - 1/2)] \ . \qquad (1)$$

Here the theta function is equal to zero if its argument is non-positive and unity otherwise. This map is shown in Figure 1a and b.

There are two drawbacks to this particular invertible map. (1) It is not area-contracting, so no attractor can form (only something like a "weak" analogue). (2) It is not differentiable everywhere: There is the "knife" operator involved (the theta function). The first shortcoming is easy to amend (see Figure 1c). All one needs to assume is that the dough, while being squeezed flat by the rolling pin operator, weeps out part of the fluid matrix (leavening). "Contracting baker's transformations" were recently considered independently by several authors [13-15]. The simplest explicit version appears to be:

$$x_{n+1} = 2x_n \qquad (x \bmod 1)$$

$$y_{n+1} = (1/3)y_n + (2/3)\Theta(x_n - 1/2) \ . \qquad (2)$$

Here x is assumed to be confined to a circle (mod 1) for simplicity.

The second drawback mentioned above is more serious if one wants to "think simple". Indeed Smale [16], who was among the first to look at simple (if possible linear) differentiable systems showing nontrivial dynamical behaviour, apparently already tried hard to find an embedding two-dimensional manifold on which the two ends of the present x-ring can be made flush in such a way that internal pieces fit together smoothly through all iterates. A two-dimensional area-contracting linear map with this desirable property would be a nice object to study. Unfortunately, the second dimension as it is given to us in this world somehow spoils this game. One has to move up to three dimensions if one wants to have all these tasty properties (linearity, volume contraction, differentiability) combined.

You may still feel like not giving up and try embedding on a two-torus with a slit. This attempt is illustrated in Figure 1d. The closed strip on the left (zeroth iterate) can indeed be embedded, after a two-fold elongation accompanied by sufficient vertical shrinking, into the torus to the right. Similarly at the next step: The again length-doubled strip can be embedded into an even longer (now wrapped up) torus with a slit again. (And so forth). A torus with a slit, however, is nothing but an ordinary strip. The only difference to the original closed strip (first picture in Figure 1e) is that the right-hand margin and the left-hand one are now merged in a "crossed over" fashion (see the second picture in Figure 1e). The series of pictures in Figure 1e is, therefore, an illustration of what one would like to have in two dimensions. Only one cannot quite get it since the present sequence of maps is <u>not</u> described by a (single) diffeomorphism. The domain of the map has to be subtly re-defined at every step (crossing-over of two more ends).

This explains, perhaps, why Smale [16] came up with his "solenoid"-generating three-dimensional map (Figure 1f). This time it is a solid torus that is stretched and squeezed and wrapped up once before being put back into itself. (Actually, it could have been a hollow solid ring from the beginning since there is an unused ring-shaped core in the middle of the structure that could have been left out). An explicit difference equation in three variables (with mixed Cartesian and polar coordinates) is available for the map of Figure 1f [17]. A 4-variable ordinary differential equation believed to posses a map of the present type as a Poincaré cross section also exists [18].

Still, much of the intuitive appeal that a two-dimensional system (like that of Figure 1e) would have possessed if it worked, is gone with the transition to the third dimension. Therefore it is reassuring that a 2-dimensional contracting Axiom-A map (this technical term for the present class of systems is defined in [16]) indeed exists. Plykin [19] succeeded in finding a two-dimensional example (incidentally non-circular) which possesses all the desired properties except for linearity. Unfortunately, this map is so complicated (involving 9 fixed points of rigidly prescribed properties and mutual orientation; see [20] for a re-drawing in the convention of horseshoe-shaped submaps) that there is little hope that an explicit (not to say, simple) equation for this map will be found soon.

All of this suggests the following easy way out. Let us simply pretend that the map of Figure 1e did work as an embedding of Eq.(2) (and forget about the fact that in reality a re-definition of some border identities will be required at every step). This fictitious linear map then constitutes a prototype example worth to be considered when invertible systems with chaos have to be studied from a very general point of view - as when the effects of infinite iteration indices are at stake (see the next Section).

Of course it would also be possible to use the solenoid-generating map instead. But there is another disadvantage to that map yet to be mentioned. If results of maximum generality in the dimension in question are desired, it is necessary to change the map slightly (by letting the first iterate have 3 loops rather than two). The map in this case generates "noodle-map chaos" (one of the generic types of behaviour possible in three dimensions only but nevertheless analogous to ordinary chaos in two [14]). Even though there is still only one direction of repetitive stretching (one positive Lyapunov characteristic exponent) present, the complexity of the dynamical behaviour occurring is vastly increased. This follows from the fact that the time-reversed map now possesses two nontrivial positive Lyapunov characteristic exponents, cf.[14]. A simple piecewise-linear map illustrating these points is available [21].

Therefore, it is fitting to add one last excuse for the above proposal to stick to Eq.(2) "embedded in Figure 1e". There is a proof that a map like this indeed exists. An area-preserving linear differentiable invertible map of a two-torus exists (the well known Sinai-Thom-Anosov diffeomorphism [22], cf.[16]) that is so well behaved ("structurally stable" [23]) that an area-contracting version arbitrarily close to it, retaining its qualitative features, is bound to exist. Unfortunately, an explicit equation for this derived map [16,24] is not available. It is therefore not known to date whether or not a linear equation as simple as Eq.(2) (note that putting Eq.2 on the right manifold would make it linear) can describe a two-dimensional map with an Axiom-A attractor. This is a purely quantitative problem, however. Qualitatively, the above model system appears to be admissible.

4. A LONG-LINE ATTRACTOR

The mathematical properties of the above-proposed invertible, differentiable map (Eq. 2 embedded into Figure 1e) are rather well known. This is because the present map is nothing but an especially simple example of an Axiom A diffeomorphism in the sense of Smale [16]. Specifically, it is known that for iteration index n going to infinity, an Axiom-A attractor Λ is formed. Λ is an invariant set ($\Lambda^+ = \Lambda^-$), and the periodic points (countably many) are dense on it. These are perhaps its two most important properties (cf.[16] and [25]).

These well known results retain their significance if higher (transfinite) iteration numbers are admitted. There remains an invariant subset (L) in existence. On L, the periodic points (up to countable periodicity) are dense. The set L consists of all these periodic points (as mentioned) plus their associated unstable manifolds. By formally introducing the closure of L (for the case n goes to infinity), one re-obtains the attractor Λ [26].

However, this last step is no longer necessary now - and indeed would be unduly restrictive. It is still true that asymptotically (for $n \to \infty$), all initial points come arbitrarily close to L. (This was the motivation for defining the closure of L to be the attractor). But it could also be the case now that another attracting set of measure zero in the domain of attraction exists that is so much larger than L that L forms only a measure-zero "skeleton" in it. This larger set L would be analogous to the closure of L (that is, Λ), but would not necessarily be identical to it. Specifically, it need not be invariant.

The postulated attracting set L indeed exists. It is a long line in the sense of topology (see [27]) as is easy to see. Assume $n = \aleph_0$ (countably infinite). Then a vertical cross section through the map of Figure 1e or Figure 1c (see the dotted line in Figure 1c, right-hand picture) yields a standard Cantor set of lines. These uncountably many lines of equal lengths are, by virtue of the properties of the successive toroidal embedding manifolds assumed (Figure 1e), all mutually connected in such a way as to form a single closed line. This single attracting closed line is a long line. This is because it could have been obtained also by choosing all ordinals up to the least uncountable one and then inserting a unit interval in between every adjacent pair. This is the definition of the long line as given in [28].

The long line is both non-compact and non-separable [28]. The former property means that not every covering has a finite (or even countable) subcover. The second means that no countable set (like a periodic orbit of at most countable periodicity) can be dense on it.

The latter feature could be formally amended by admitting periodic solutions of transfinite periodicity (whose existence was shown in Section 3) into the non-wandering set. A periodic orbit could then, once more, be said to be "dense" on the attractor.

The former feature is more incisive. It means that the attractor is not an invariant set. This can be seen more directly by focusing on the property of "bottomlessness" (unfinishedness-in-principle) of the present attracting object. Note that for iteration index $n = \aleph_1$, the attracting long line assumes the length of $\aleph_2$ unit intervals, and so forth. L therefore is not just one long line, but the set of all possible long lines. Moreover, due to the transfinite accuracy (being infinitely fine not only asymptotically,but actually) assumed, almost all points on the never-finished attracting long line remain bijectively related to their first preimage in the rectangular domain of attraction. An exception is only the zero-measure subset of the domain of attraction that is formed by the stable manifolds of the periodic points - since it is attracted by the invariant set L. The set *L*-L can, therefore, be said to constitute an example of a "transfinitely unwrappable" set.

5. DISCUSSION

Chaos and transfinite accuracy are two intimately linked concepts. This may be the ultimate reason for the intuitive appeal of the phenomenon of chaos. To keep track of position in spite of mixing demands believing in the impossible.

Apparently, Anaxagoras [29] was the first to be intrigued by this idea. He effectively changed the meaning of the word Chaos from "bottomless abyss" (châo means "I yawn" and chaos in this sense was the first primordial entity in Hesiod's mythology) to "perfect mixture". While there is no historical proof that Anaxagoras himself attached the old name to his own new primordial entity (there are only 4 pages of his textbook surviving), he certainly endowed that latter with the main feature of the former (bottomlessness).

In Fragment 12, he says [29] that even though the mixture had existed in an invariant state for an infintely long time, it eventually got unwrapped by a recurrent motion. This "recurrent motion" is mentioned 8 times. Instead of Birkhoff's Latin term, a Greek word, "perichoresis" (moving around), was used. The author was careful to make it clear to the reader it was not a repetitive (periodic) motion that he had in mind but one whose influence changed and grew so that it eventually succeeded in unmixing the whole primordial mixture. In other words, Anaxagoras described an "unmixing machine" (see [30]).

The theory of physical mixing was, in those days, apparently not unfamiliar to many thinkers. The production of butter out of milk, and of sausage out of another life-giving fluid, was a major conceptual paradigm. Compare in this context one of the stories on Hun Tun (Chaos) in Chinese mythology. The primordial entity here is a leather bag filled with blood and there is a fiery arrow mentioned in connection with it [31]. In India, there is the picture of Vishnu lying at the bottom of a sea of milk that he slowly churns (unmixes?) with a rod as big as a mountain (see [32] for an ancient drawing).

The fact that the idea of an unmixing process after infinite time is not unproblematical mathematically was apparently seen by Anaxagoras already. After having stressed that everything was completely intermixed in the eternal primordial state, he unexpectedly was able to give a reason why and how the unmixing motion could have got started at one point in space and time: There was a single substance left that had been too "fine" to be miscible, the mind (Noûs) [33].

Anaxagoras' proposal to attribute a transfinite accuracy (cf. also Fragments 3 and 5; [29], p.370) to the mind, in the context of mixing, is indeed paradoxical. Returning to topology, we find ourselves confronted with the following dilemma: Every successive iterate of the map is contained, as a closed subset, in the preceding one. The length of each successive image set also is just twice that of the pre-image into which it is embedded. In other words, the finite intersection condition (cf.[34]) is always fulfilled over any finite number of steps. Therefore, there ought to ultimately exist a compact invariant object (the "last iterate")[35].

Rather than suggesting an easy solution (add "ultimate nondivergence" to the conditions under which nested closed sets have a compact intersection), it is perhaps fairer to leave this problem completely open.

To conclude, the generalized Peano curves [14] (or fractals [36]) that are formed as attractors under the iteration of chaos-producing invertible maps of two and more dimensions apparently belong to a very special type of such objects. They are: The long line, the big sheet, the great cloud, ...

6. SUMMARY

Chaotic attractors in invertible systems are fractals of the long-line type. The contracting baker's transformation without cuts (an idealized 2-dimensional invertible map of Axiom-A type) provides an example. The long-line attractor appears only if infinite iteration indices are admitted. The naturalness of this latter assumption follows from iteration theory. Its necessity follows from a reconsideration of the Sharkovsky point in the parameter space of the logistic recurrence: At this point, a periodic attractor of "uncountable periodicty" can be shown to exist. If an infinite iteration index is formally admitted, the attractor of the invertible map proposed consists of uncountably many unit-length segments joined together. It in consequence forms a long line in the sense of topology. A long-line attractor is both non-compact (non-invariant) and non-separable (a periodic orbit of at most countable periodicity cannot be dense on it). It constitutes an example of a transfinitely unwrappable set in the sense of Anaxagoras.

ACKNOWLEDGEMENTS

I thank György Targonski for having brought to my attention the literature on the long line. I also thank him and an anonymous referee for many helpful suggestions. A discussion with H.E.Nusse is also gratefully acknowledged.

FIGURE CAPTION

Fig. 1. The baker's transformation (Eq.1), shown in detail (a) and for short (b). A contracting baker's transformation (Eq.2), shown in the same manner (c). A proposed "differentiable contracting baker's transformation" (e), making use of the two different possible ways to close a ring shown in (d). (Letters stand for gluing-together). The question whether or not a two-variable equation may be found for (e) is dealt with in the text. In (f) a t h r e e - dimensional differentiable contracting analogue, Smale's solenoid [16], is reproduced.

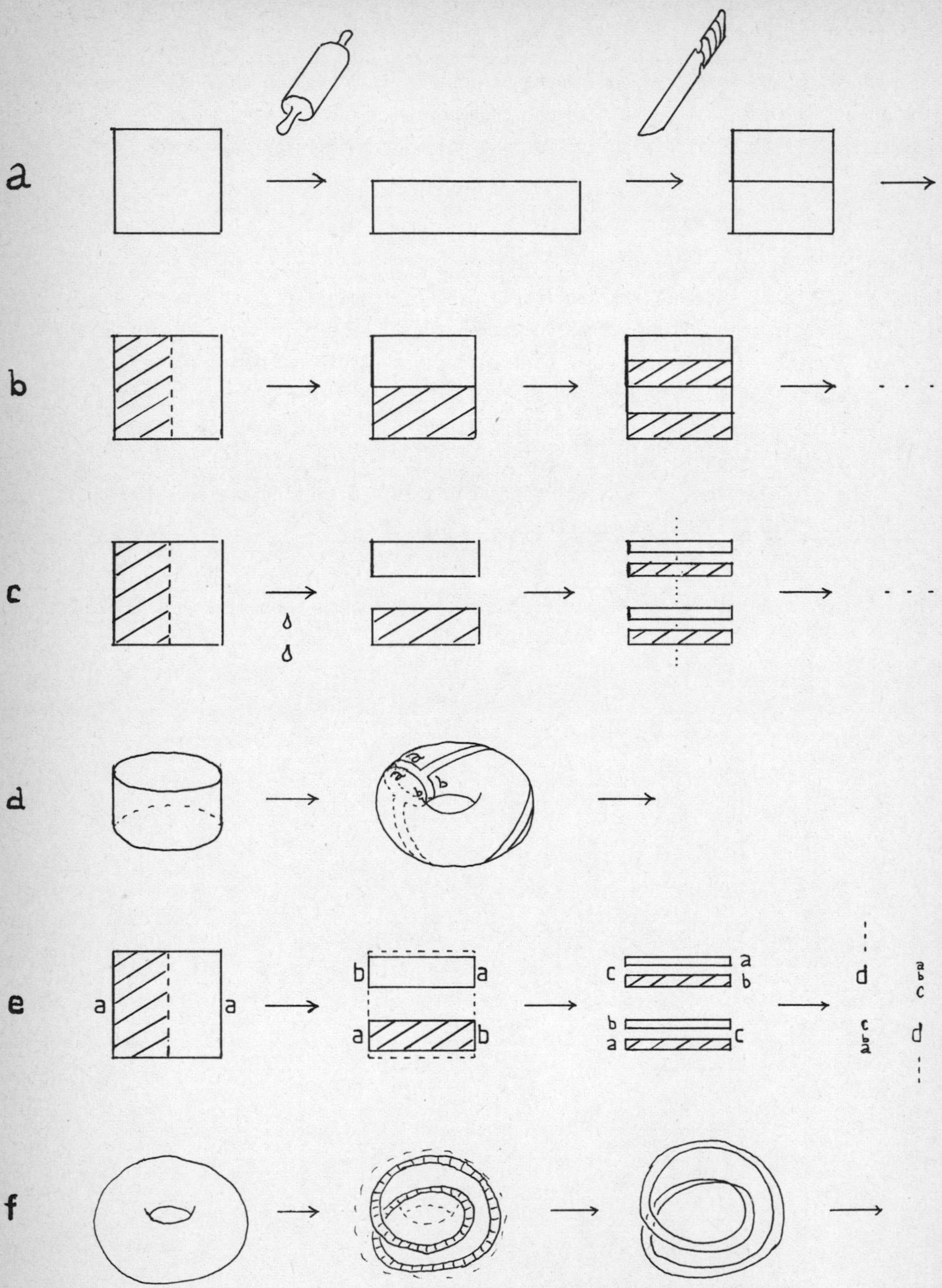

Fig. 1

REFERENCES

[1] Introduction. These Proceedings.

[2] Targonski, Gy.: "Topics in Iteration Theory". Göttingen, Vandenhoek and Ruprecht 1981.

[3] Sharkovsky, A.N.: Co-existence of cycles of a continuous map of a line into itself. Ukr. Math. Z. 16, 61-71 (1964).

[4] Feigenbaum, M.J.: Quantitative universality for a class of nonlinear transformations. J. Stat. Phys. 19, 25-52 (1978).

[5] Rössler, O.E.: Syncope implies chaos in walking-stick maps. Z. Naturforsch.32a, 607-613 (1977).

[6] Grossmann, S., and Thomae, S.: Invariant distribution and stationary correlation functions of one-dimensional discrete processes. Z. Naturforsch. 32a, 1353-1363 (1977).

[7] Lorenz, E.N.: Noisy periodicity and reverse bifurcation. Ann. N.Y. Acad. Sci. 357, 282-291 (1980).

[8] Misiurewicz, M.: Absolutely continuous measures for certain types of maps of an interval. Publ. Math. I.H.E.S. 53, 17-51 (1981).

[9] Daido, H.: Universal relation of a band-splitting sequence to a preceding period-doubling one. Phys. Lett. 86 A, 259-262 (1981).

[10] Li, T.Y., and Yorke, J.A.: Period three implies chaos. Am. Math. Mon. 82, 985-992 (1975).

[11] Thom, R.: "Stabilité Structurelle et Morphogénèse. New York. Benjamin 1972.

[12] Hopf, E.: "Ergodentheorie". Berlin. Springer-Verlag 1937, p.42.

[13] Kaplan, J.L., and Yorke, J.A.: Chaotic behaviour of multidimensional difference equations. Springer Lect. Notes Math. 730, 228-237 (1979).

[14] Rössler, O.E.: Chaos and bijections across dimensions. In "New Approaches to Nonlinear Problems in Dynamics".(P.J.Holmes, ed.), 477-486, Philadelphia, SIAM 1980.

[15] Aizawa, Y., and Murakami, C.: Generalization of baker's transformation - chaos and stochastic process on a Smale horseshoe. Progr. Theor. Phys. 69, 1416-1426 (1983).

[16] Smale, S.: Differentiable dynamical systems. Bull. Am. Math. Soc. 73, 747-817 (1967).

[17] Smale, S.: Dynamical systems and turbulence. Springer Lect. Notes Math. 615, 48-70 (1977).

[18] Rössler, O.E.: Example of an Axiom-A O.D.E. In "Chaos, Fractals and Dynamics". (P.Fischer and W.R.Smith, eds.), 105-114. New York. Marcel Dekker 1985.

[19] Plykin, R.V.: Sources and sinks of A-diffeomorphisms of surfaces. Math. USSR Sbornik 23, 233-253 (1974).

[20] Rössler, O.E.: Chaos and strange attractors in chemical kinetics. In "Synergetics far from Equilibrium". (A.Pacault and C.Vidal, eds.), 107-113. Berlin-New York. Springer-Verlag 1979.

[21] Hudson, J.L., and Rössler, O.E.: A piecewise-linear invertible noodle map. Physica 11 D, 239-242 (1984).

[22] Sinai, Ja.: On the concept of entropy for a dynamical system (in Russian). Dokl. Akad. Nauk SSSR 124, 768-771 (1959).

[23] Anosov, D.V.: Roughness of geodesic flows on compact Riemannian surfaces of negative curvature. Soviet Math. Dokl. 3, 1068-1070 (1962).

[24] Williams, R.F.: The 'DA' maps of Smale and structural stability. In "Global Analysis".(S.S. Chern and S. Smale, eds.), Proc. Symp. Pure Math. (AMS) 14, 329-334 (1970).

[25] Nusse, H.E.: "Chaos, yet No Chance to Get Lost". Utrecht. Elinkwijk 1983.

[26] Williams, R.F.: One-dimensional nonwandering sets. Topology 6, 473-487 (1967).

[27] Alexandroff, P.S., and Urysohn, P.: Mémoire sur les surfaces topologiques compacts. Verh. Konink. Acad. Wetensch. Amsterdam 14, 1-96 (1929).

[28] Steen, L.A., and Seebach, J.A.: "Counterexamples in Topology", 2nd edn. New York. Springer-Verlag 1978, p.71.

[29] Anaxagoras of Clazomenae: "Physis". Fragment 12. ($\sim$ 462 B.C.) Reprinted and translated in: G.S.Kirk and J.E.Raven, "The Presocratic Philosophers, A Critical History with a Selection of Texts". Cambridge. Cambridge University Press, 372-373 (1957).

[30] Rössler, O.E.: The chaotic hierarchy. Z. Naturforsch. 38a, 788-801 (1983).

[31] Christie, A.: "Chinese Mythology". London. Paul Hamlyn, p.44 (1968).

[32] "Nova - Adventures in Science". New York. Addison-Wesley, p. 27 (1982).

[33] Note the arrow-like quality of this dual second principle again.

[34] Munkres, J.R.: "Topology - A First Course". Englewood Cliffs, N.J. Prentice-Hall 1975.

[35] Note that this is how Smale [16] originally introduced his Axiom-A attractor: as the intersection of all ($n > 0$) iterates of its domain of attraction, yielding a compact invariant set Λ. (He then went on to characterize the object further as being locally the product of a Cantor set and a one-dimensional arc, in the simplest case [17]. In the present terminology, this would amount to postulating the existence of a "compact long line", provided such were possible).

[36] Mandelbrot, B.: "The Fractal Geometry of Nature". San Francisco. Freeman 1982.

O.E. Rössler
Universität Tübingen
D-7400 Tübingen, FRG

PROPERTIES OF INVARIANT CURVES NEAR A KNOWN INVARIANT CURVE

A. Sabri

INTRODUCTION

It is well known [1] that the global solution structure of a real-valued equation like.

$$\ddot{x} + \omega_0^2(1+h\cos\omega t)\sin x = 0, \quad x = x(t), \quad \cdot = d/dt, \tag{1}$$

is completely defined by the properties of the associate recurrence

$$x_{n+1} = f(x_n,y_n), \quad y_{n+1} = g(x_n,y_n), \quad y_n = \dot{x}_n, \tag{2}$$

where $n = 0, \pm1, \pm2,\ldots,x_n = x(t_n)$, $t_n = nT$, $T = 2\pi/\omega$. Since equation (1) is Hamiltonian, it has been conjuctured [2] that (2) can be transformed to the form

$$\xi_{n+1} = \eta_n + F(\xi_n), \quad \eta_{n+1} = -\xi_n + F(\xi_{n+1}) \ . \tag{3}$$

This conjecture has been confirmed by accuracy controlled numerical computations, F turns out to be periodic of period 2π and its form is shown in Fig.1 for two sets of parameter values. For $h \neq 0$ the recurrence (2) or (3) can be used to construct a phase portrait of (1) in the "discrete" phase plane x_n, y_n, which coincides by definition with that in the "continuous" phase plane $x(t)$, $y(t)$ for $h = 0$ [3]. An example of the latter is shown in Fig.2. The trajectories of Fig.2 turn into invariant curve C: $W(x_n,y_n) = C$, where $W(x,y)$ is a suitably chosen smooth solution [2] of the functional equation

$$W(f(x,y),g(x,y)) = W(x,y) \ . \tag{4}$$

Segments of C passing through non-degenerate saddles of (2) can be described analytically by means of convergent series [4], and then continued by iteration. Convergent series are also known for segments of C passing through nodes or meeting at a cusp [2]. Near a singular point of (2) an approximate local phase portrait of (1) can be constructed directly by two types of asymptotic series [5], [6]. This local phase portrait provides a one-parameter family of solution of (4) near a particular analytically known invariant curve C_0 : $W(x,y) = C_0$. Since the quality of the approximation is not known a priori, its accuracy is checked by means of an independent numerical computation, carried out for a certain number of values of C close to C_0.

TYPICAL FEATURES OF THE INVARIANT CURVE STRUCTURE

The starting point for the determination of the invariant curves C is the knowledge of point-singularities of (2), i.e. of fixed points and cycles of order $k > 1$. In mechanics the former correspond to "main" resonances of (1), and the latter to "subharmonic" ones. It is well known that (1) admits and infinity of resonances. The most important ones for mechanical oscillations [7] are shown in Fig.3 and 4, together with a part of the invariant curves passing through the "main" saddles ($x_n = \pm\pi$, $y_n = 0$). In the notation k/r, k designates the order and r the "rotation number" of the resonance. When $h \neq 0$, the interior region still exists but it is smaller in size and has a much more complicated structure. The rough features of the local phase portrait near the lowest order resonances are shown in Fig.5 to 10 [7]. The 'fine structure' induced by high order resonances with large rotation numbers has been smoothed out for more clarity; it is unimportant for mechanical applications. In a theoretical study it would constitute the main topic of interest, because it is of the box-within-the-box type [2], i.e. it is of "fractal" nature. This fine structure appears to constitute the main obstacle in the direct analytical study of the functional equation (4), even if it is restricted to a small neighbourhood of a low order resonance.

When $h = 0$, a description of 'rotation invariant curves' verifies the functional equations

$$x(t+\tau) = x(t) + 2\pi, \qquad \dot{x}(t+\tau) = \dot{x}(t) , \tag{5}$$

where τ is a known constant (time needed for one rotation). The value of τ can be expressed in terms of x_0, y_0, ω_0 by means of an elliptic integral. What happens to these invariant curves when $h \neq 0$, is still a matter of speculation. Exploratory analytical and numerical studies indicate that if (5) is assumed to hold, then can no longer be a constant. Two typical forms of $\dot{x}(t)$vs $x(t)$, determined numerically are shown in Fig.11. They do not resemble anyhting observed so far in the oscillation region of the phase portrait. No trace could be found in published literature of an analytic method suitable for their analytical study.

FIGURES

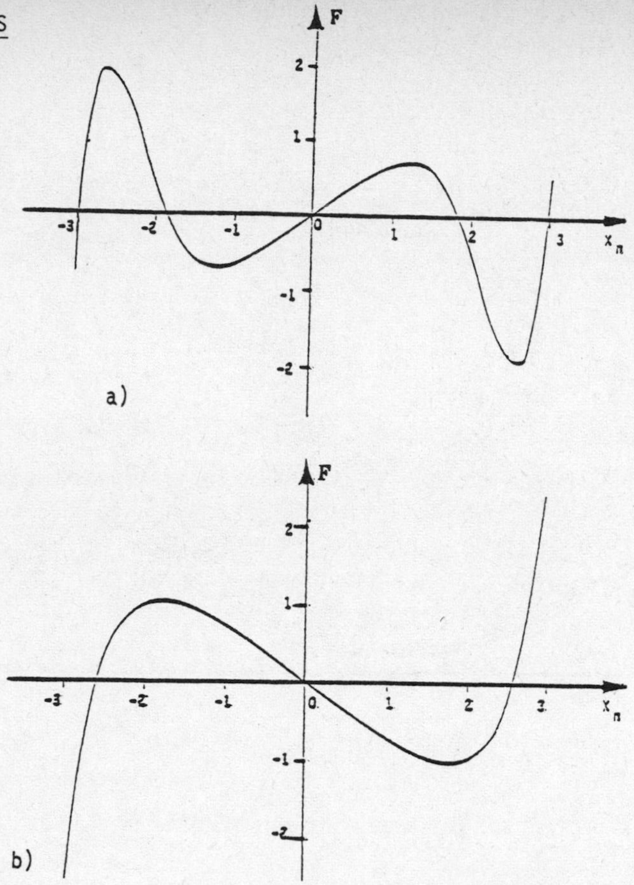

Fig. 1. Characteristic function of the associate recurrence of equation (1):

a) $h = 0.001$, $\omega_0^2 = 4.01$, $\omega = 2$.

b) $h = 0.01$, $\omega_0^2 = 1.01$, $\omega = 2$.

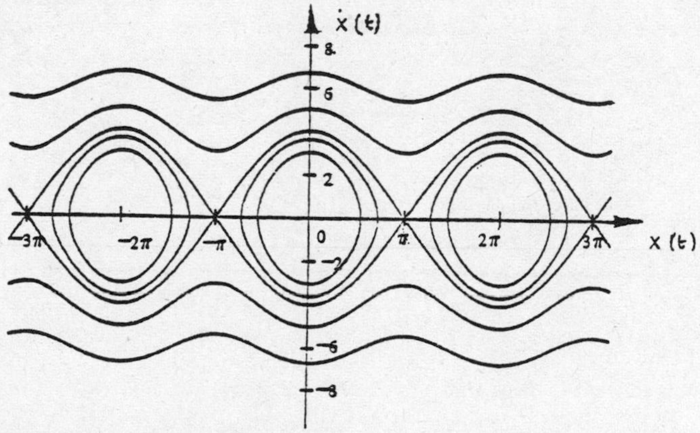

Fig. 2. Phase plane, $h = 0$, $\omega_0^2 = 1$, $\omega = 2$.

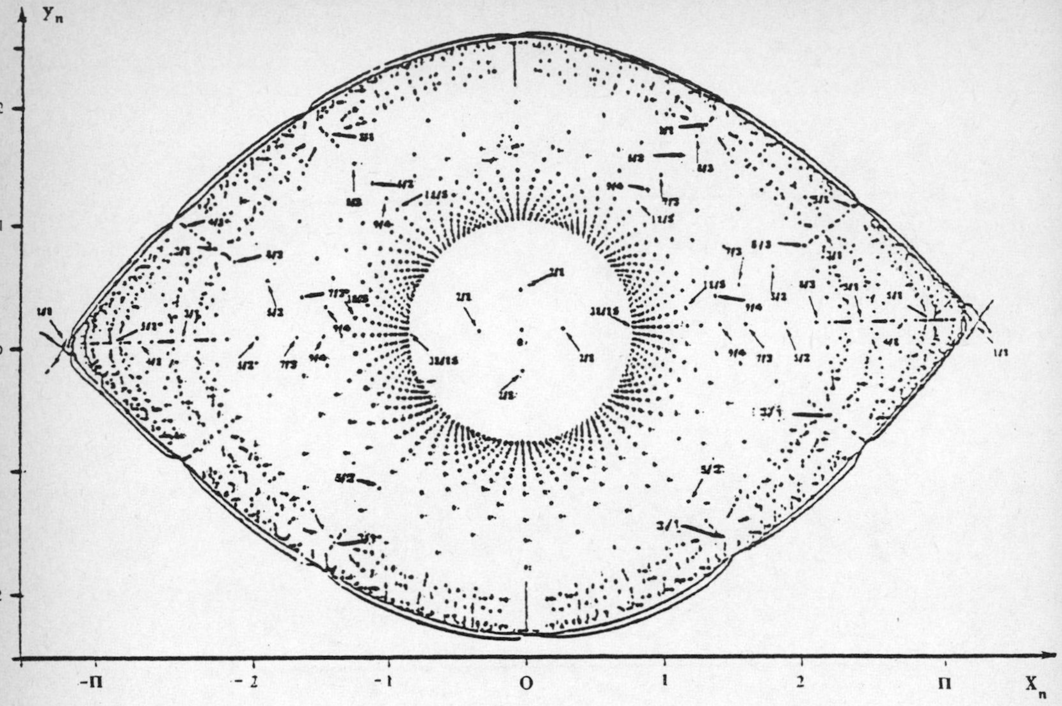

Fig. 3. Main invariant curves and distribution of cycles.
$h = 0.01,\ \omega_0^2 = 1.01,\ \omega = 2,\ S1 = 23.50.$

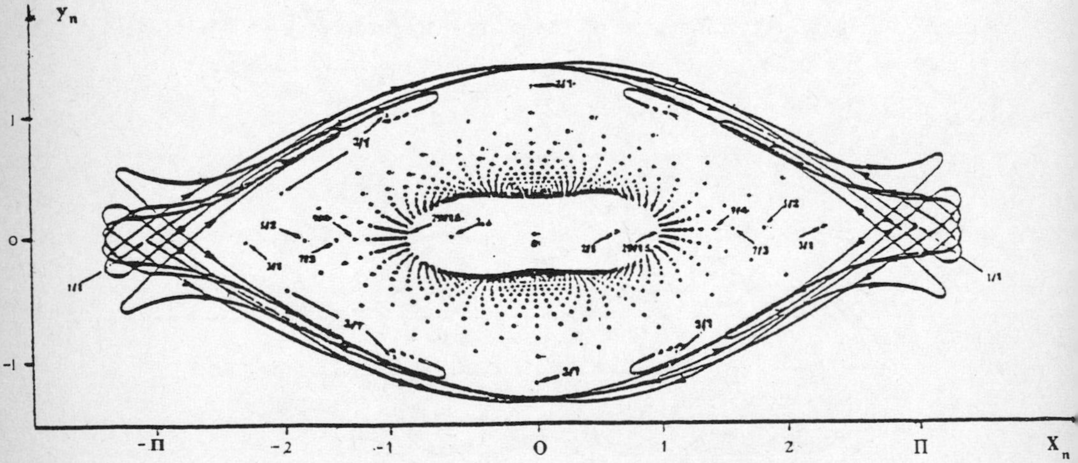

Fig. 4. Main invariant curves and distribution of cycles.
$h = 0.11,\ \omega_0^2 = 1.01,\ \omega = 2,\ S1 = 23.67.$

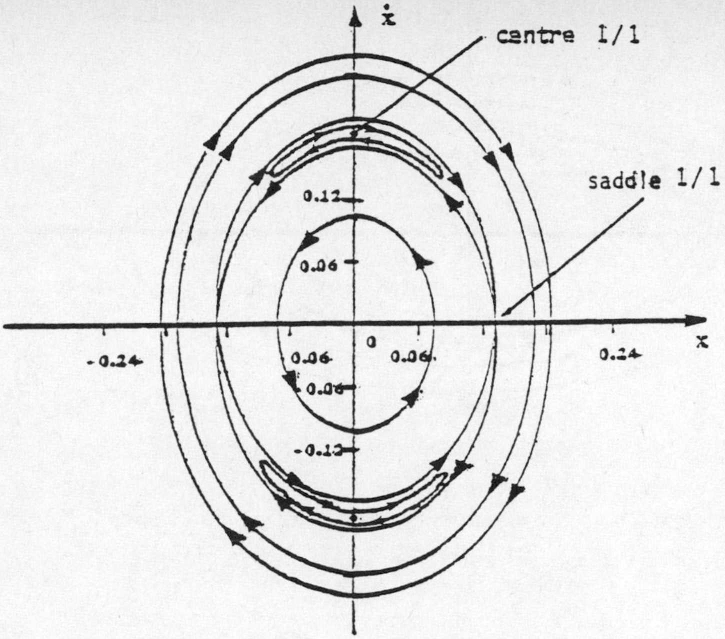

Fig. 5. Approximate invariant curves (determined by the Von Zeipel method) near the main resonance:
$\omega^2 \cong \omega_0^2 = 4.01, \quad h = 0.01, \quad \omega = 2.$

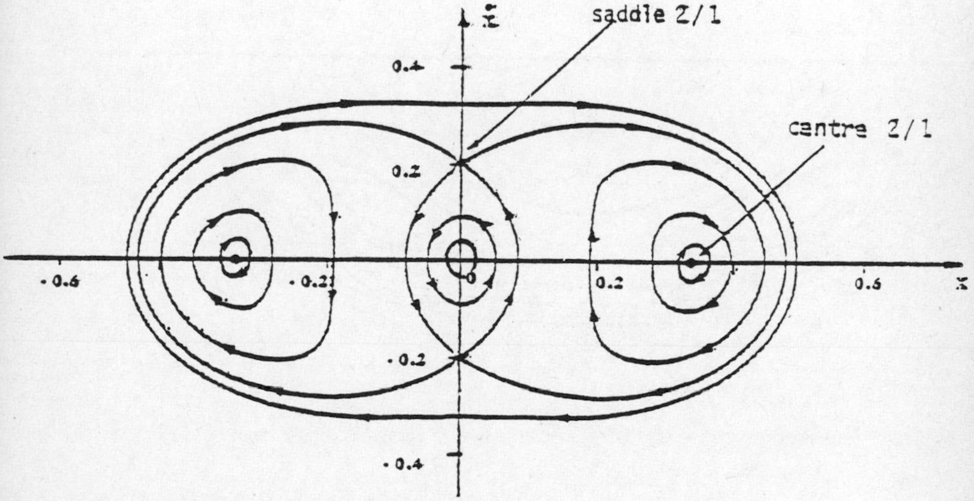

Fig. 6. Approximate invariant curves near the sub-harmonic resonance of order two:
$(\frac{\omega}{2})^2 \cong \omega_0^2 = 1.01, \quad h = 0.01, \quad \omega = 2.$

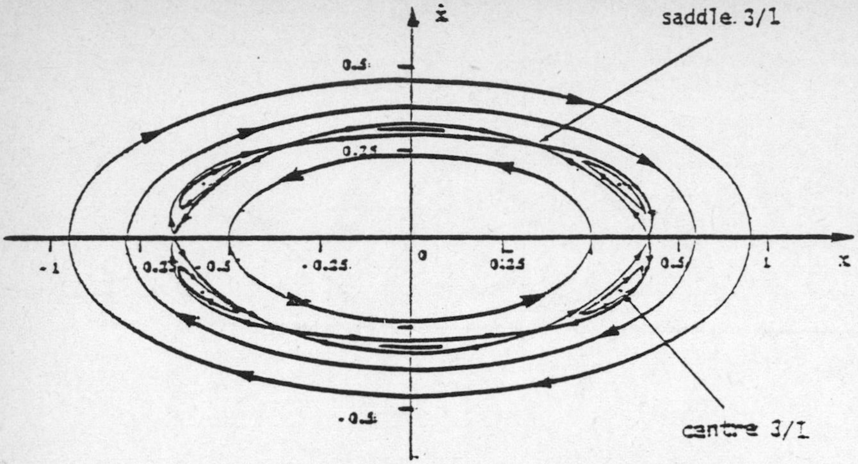

Fig. 7. Invariant curves near the sub-harmonic resonance of order three: $(\frac{\omega}{3})^2 \cong \omega_0^2 = 0.48, \quad h = 0.1, \quad \omega = 2.$

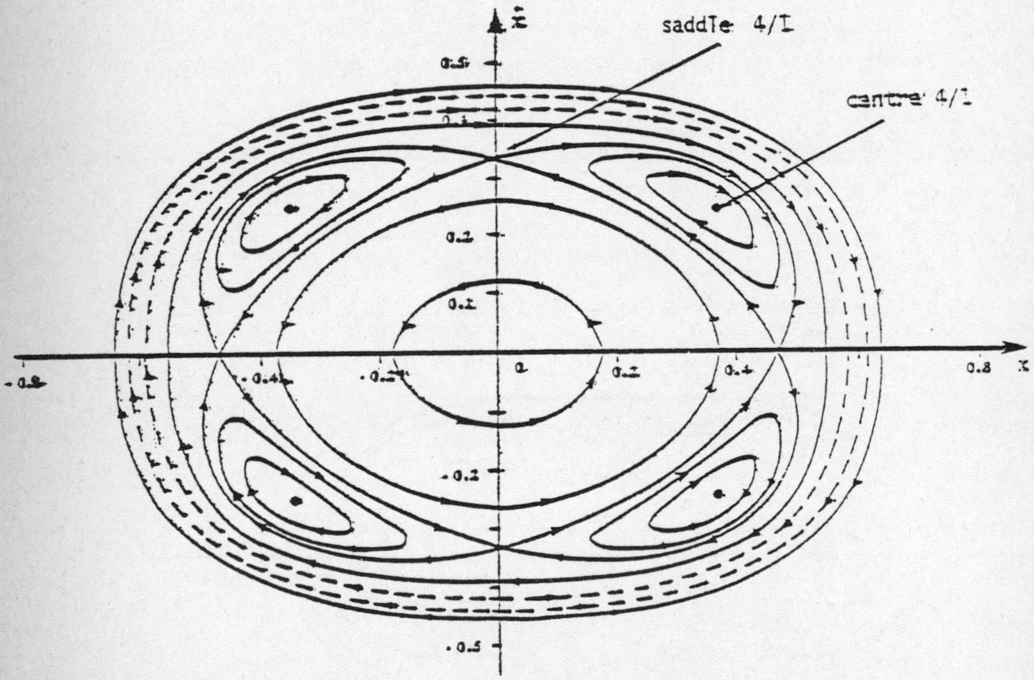

Fig. 8. Invariant curves near the sub-harmonic resonance of order four: $(\frac{\omega}{4})^2 \cong \omega_0^2 = 0.26, \quad h = 0.36, \quad \omega = 2.$

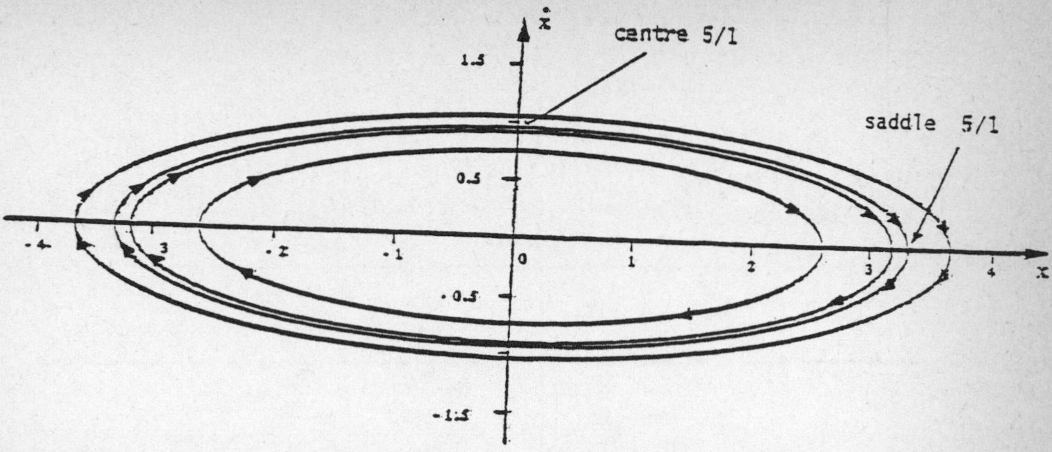

Fig. 9. Invariant curves near the sub-harmonic resonance of order five:
$(\frac{\omega}{5})^2 \simeq \omega_0^2 = 0.17$, $h = 1$, $\omega = 2$.
Cycle points are not resolved at the scale used.

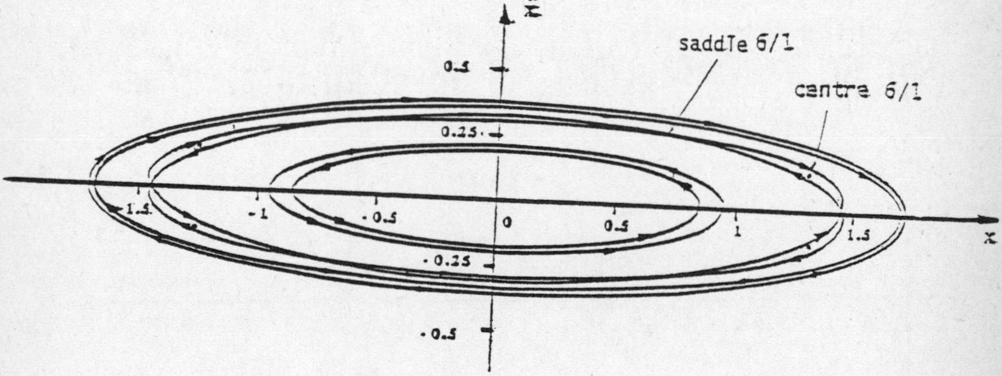

Fig.10. Invariant curves near the sub-harmonic resonance of order six:
$(\frac{\omega}{6})^2 \simeq \omega_0^2 = 0.18$, $h = 0.01$, $\omega = 2$.

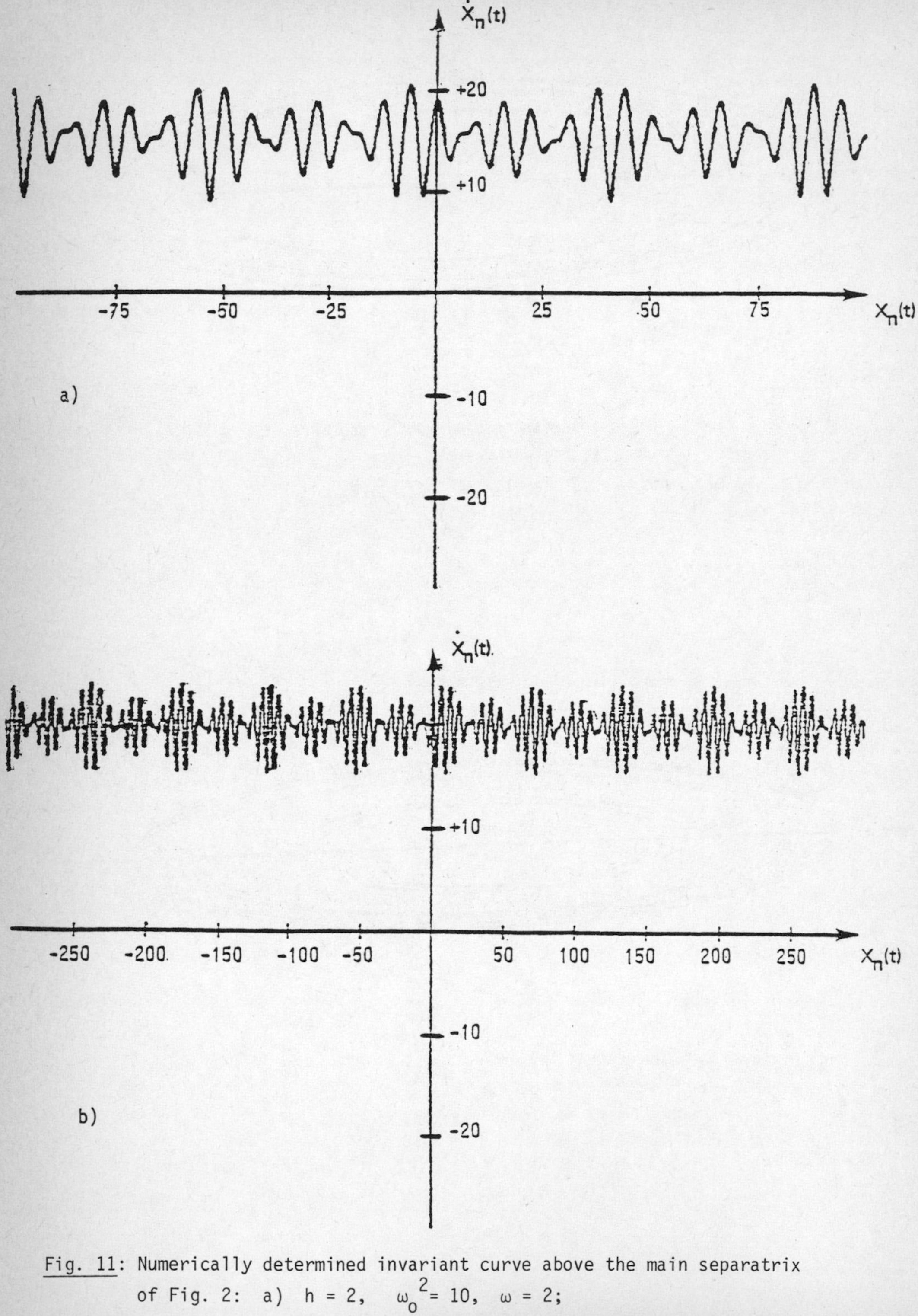

Fig. 11: Numerically determined invariant curve above the main separatrix of Fig. 2: a) $h = 2, \quad \omega_0^2 = 10, \quad \omega = 2$;

b) $h = 5, \quad \omega_0^2 = 14, \quad \omega = 2$;

REFERENCES

[1] Poincare, H.: "Les mêthodes nouvelles de la mêchanique cêleste", Vol.I,II,III, Gauthiers-Villars, (1892, 1893, 1899).

[2] Gumowski, I, Mira, C.: "Recurrences and discrete dynamic systems", LN in Math., Springer, n° 809, Berlin, (1980).

Gumowski, I, Mira, C.: "Dynamique chaotique, transition ordre desordre", Cêpadues, Toulouse, (1980).

[3] Gumowski, I.: "Some relations between differential equations, recurrences and functional iterates", Actes, I.C.N.O., v.2, Naukova Dumka, Kiev, (1984), p.98-103.

[4] Lattes, S.: "Sur les êquations fonctionnelles dêfinissant une courbe ou une surface invariante par une transformation", Ann. di mathematica, serie 3, vol.13, (1906), p.1-69.

[5] Zeipel, H.V.: "Recherche sur le mouvement des petites planêtes", Arkiv. för Math., Astron., Physik, vol.11, n° 1 et 7, (1916), p.1-58, et p.1-62.

[6] Bogoliubov, N., Mitropolsky, I.: "Asymptotic methods in the theory of non linear oscillations", GOS. Izd. Fiz. Lit., Moscow, (1955).

[7] Sabri, A.: "Etude des rêsonances paramêtriques d'un systême dynamique non linêaire â pêriodicitê spatiale", Thêse U.P.S. n° 2818, Toulouse (1983).

U.E.R. de mathêmatique
Universitê PauL Sabatier
118 route de Narbonne
F-31077 Toulouse Cedex

NORMAL FORMS FOR SYSTEMS OF FORMAL POWER SERIES COMMUTING IN PAIRS AND ITERATION PROBLEMS

Jens Schwaiger

Starting with well known results (see e.g. [1], [2], [3]) and motivated by concepts taken from [4] we give a new interpretation for normal form problems for formal power series. After this a theorem on formal power series commuting in pairs is proved, which enables us to generalize a theorem of Reich and Bucher ([5],[6]) on continuous iterations. Additionally we get a necessary condition for the general iteration problem without regularity conditions. Finally I want to mention C. Praagman's paper [8] containing results partially equivalent to ours, but using completely different methods, and to say, that an enlarged version of this paper containing full proofs will appear elsewhere.

§1. INTRODUCTION

Concerning notions connected with formal power series the reader is asked to consult, say, [2].

For short I write Ω for the set of those formal power series (= f.p.s.) $F \in (\mathbb{C}[\![X]\!])^n$ fulfilling $F(0) = 0$.

Furthermore, for $t_o = 0 < t_1 < \dots < t_r = n$ and $\tau = (t_1,\dots,t_r)$, a matrix $A \in M_n(\mathbb{C})$ is called to be of type τ, if A is of the form

$$A = A_1 \oplus A_2 \oplus \dots \oplus A_r \tag{1}$$

where $A_i \in M_{s_i}(\mathbb{C})$, $s_i := t_i - t_{i-1}$.

A matrix A of type τ is called to be a special triangular matrix of type τ, if in (1) we have $A_i = \rho_i E_i + N_i$, E_i the unit matrix in $M_{s_i}(\mathbb{C})$ and N_i some nilpotent upper triangular matrix in that space of matrices. $d(A) := \overset{r}{\underset{i=1}{+}} \rho_i E_i$ in called the diagonal part of A. For the set of special triangular matrices of type τ I write $\Delta_{n,\tau}$. $\rho(A) := (\rho_1,\dots,\rho_r)$ is the vector of the eigenvalues ρ_i. The following theorem (Jacobson [9], p.134) is used frequently.

Theorem 1. Let $\mathfrak{A} \subseteq M_n(\mathbb{C})$ be a set of matrices commuting in pairs. Then there exist integers $0 = t_o < t_1 < \dots < t_r = n$ and there exists some regular matrix T, such that $T\mathfrak{A}T^{-1} = \{TAT^{-1} | A \in \mathfrak{A}\}$ is a subset of $\Delta_{n,\tau}$, where $\tau = (t_1,\dots,t_r)$.

Definition 1. Let $D = \bigoplus_{i=1}^{r} \rho_i E$ be a special diagonal matrix (of type τ) and let F be an element of Ω. Then F is called normalform (with respect to D), if F commutes with DX:

$$DX \circ N = N \circ DX \quad . \tag{2}$$

Remark 1. The notion of normal forms used formerly (see e.g.[2]) is a special case. Namely the case, where the linear part lin F of F is a Jordan canonical form and where D is the diagonal part of J. Those normal forms are called special normal forms from now on.

Remark 2. If F, G are normal forms with respect to D (= D - normal forms), then so is $F \circ G$ and F^{-1}, if F is invertible.

Furthermore we can rewrite and generalize a result mentioned in [2].

Remark 3. Let N be a D-normal form, $D = d(A)$, $A = \text{lin}\, N \in \Delta_{n,\tau}$. Let furthermore $M \in \Omega$ be such, that $NM = MN$ and $\text{lin}\, M = B$ is of type τ. Then M is a D-normal form.

Now, for $F \in \Omega$, write $F = \sum_{i=1}^{\infty} p_i$, where the p_i's are homeogeneous polynomials in $X = {}^t(X_1,\ldots,X_n)$ with coefficients in $\mathbb{C}^n$ of degree i and denote by $\mathcal{P}_i$ the set of those polynomials.

More general for a set $\mathcal{D}$ of diagonal matrices of type τ let us define $\mathcal{P}_k(\mathcal{D}) := \{p \in \mathcal{P}_k \mid DX \circ p = p \circ DX \ \forall\, D \in \mathcal{D}\}$.

Then we get

Remark 4. i) $\mathcal{P}_k(\mathcal{D}) \subseteq \mathcal{P}_k(\mathcal{D}')$ if $\mathcal{D}' \subseteq \mathcal{D}$

ii) $F = \sum_{k=1}^{\infty} p_k$ is a D-normal form for all $D \in \mathcal{D}$ (= a $\mathcal{D}$-normal form) if and only if $p_k \in \mathcal{P}_k(\mathcal{D})$ for all $k \in \mathbb{N}$.

Definition 2. Let B be a special triangular matrix in $\Delta_{n,\tau}$ and let k be an integer. Then $\Psi = \Psi_{B,k} : \mathcal{P}_k \to \mathcal{P}_k$ is defined by

$$\Psi(p) = B.p(x) - p \circ BX \quad . \tag{3}$$

The following holds:

Theorem 2. Let $\mathcal{D}$ be a set of diagonal matrices of type τ and let $\mathcal{P}_k' = \mathcal{P}_k(\mathcal{D})$. Furthermore let $B \in \Delta_{n,\tau}$ and $D = d(B)$.

Then: i) $\Psi_{B,k}(\mathcal{P}_k') \subseteq \mathcal{P}_k'$

ii) $\mathcal{P}_k(\mathcal{D} \cup \{D\}) = \bigcap_{j \geq 0} \ker \Psi_{B,k}^j$

iii) $\mathcal{P}'_k = \mathcal{P}_k(\mathcal{D} \cup \{D\}) \oplus \mathcal{Q}_k(\mathcal{D} \cup \{D\})$, where $\mathcal{Q}_k(\mathcal{D} \cup \{D\}) =$

$= \bigcap_{j \geq 0} \operatorname{Im} \Psi^j_{B,k}$

iv) $\mathcal{P}_k(\mathcal{D} \cup \{D\})$ and $\mathcal{Q}_k(\mathcal{D} \cup \{D\})$ are invariant under $\Psi_{B,k}$; moreover $\Psi_{B,k}\big|_{\mathcal{Q}_k(\mathcal{D} \cup \{D\})}$ is an isomorphism of $\mathcal{Q}_k(\mathcal{D} \cup \{D\})$ onto itself.

Proof. Elementary. For iii) and iv) use Fitting's decomposition (see, e.g.[10], p.335). For ii) use methods similar to those used in [2].

Later on we need

Lemma 1. If for $F = \sum_{i=1}^{\infty} f_i$, $G = \sum_{i=1}^{\infty} g_i$ the homogeneous parts f_i, g_i are elements of $\mathcal{P}_i(\mathcal{D})$ for $1 \leq i \leq i_0$, then the homogeneous parts $h_i := (F \circ G)_i$ of $F \circ G$ are elements of $\mathcal{P}_i(\mathcal{D})$ for $1 \leq i \leq i_0$, too.

§2. NORMAL FORMS FOR COMMUTING POWER SERIES

The aim of this section is to prove

Theorem 3. Let $\mathcal{F}$ be a set of invertible power series $F \in \Omega$, commuting in pairs. Then there is some type $\tau = (t_1,\ldots,t_r)$ and there is some invertible power series $S \in \Omega$, such that

1) $\operatorname{lin}(SFS^{-1}) \in \Delta_{n,\tau}$ for all $F \in \mathcal{F}$.

2) For all $D \in \mathcal{D} := \{d(\operatorname{lin}(SFS^{-1})) | F \in \mathcal{F}\}$ and all $G \in \mathcal{G} := \{SFS^{-1} | F \in \mathcal{F}\}$ we have

$$G \circ DX = DX \circ G \quad . \tag{4}$$

This theorem means-expressing it less formally-that a set of formal power series commuting in pairs can be transformed simultaneously to simultaneous normal forms. We start with

Theorem 4. Let $F = \sum_{i=1}^{\infty} f_i, f_1 = BX$ be a normal form with respect to a set of special diagonal matrices of type τ. Let furthermore B be an element of $\Delta_{n,\tau}$ and set $D_0 := d(B)$.

Then there exists an invertible formal power series $U = EX + \sum_{i=2}^{\infty} u_i$, such that U is a $\mathcal{D}$-normal form and such that additionally $U^{-1}FU$ is a normal form with respect to $\mathcal{D} \cup \{D_o\}$.

Moreover - writing $\mathcal{P}_k(\vartheta) = \mathcal{P}_k(\vartheta \cup \{D_0\}) \oplus \mathcal{Q}_k(\vartheta \cup \{D_0\})$ and $u_i = u_i^1 + u_i^2$ with $u_i^1 \in \mathcal{P}_k(\vartheta \cup \{D_0\})$ and $u_i^2 \in \mathcal{Q}_k(\vartheta \cup \{D_0\})$ - the homogeneous polynomials u_i^1 can be choosen arbitrarily.

Proof. Put $u_1 = EX$ and $U = EX + \sum_{i=2}^{\infty} u_i$ with $u_i \in \mathcal{P}_i(\vartheta)$. So by remarks 2 and 3 $N = U^{-1}FU$ is ϑ-normal form. Thus it is enough to show that, for given $u_i^1 \in \mathcal{P}_i(\vartheta \cup \{D_0\})$, $u_i^2 \in \mathcal{Q}_i(\vartheta \cup \{D_0\})$ can be choosen is such a way, that N is also a D_0-normal form.

This is done in a way similar to that used in [2] taking into account Theorem 2 and Lemma 1.

As a consequence we get

Corollary 1. If $\mathcal{P}_i(\vartheta) = \mathcal{P}_i(\vartheta \cup \{D_0\})$ for $2 \leq i \leq i_0$ then we can chose $u_i = 0$ for $2 \leq i \leq i_0$ which means that $(U^{-1}FU)_i = f_i$ for $2 \leq i \leq i_0$.

Theorem 5. Let $G_1,\ldots,G_m$ be m invertible formal power series commuting in pairs. Then there exists some type $\tau = (t_1,\ldots,t_r)$ and some invertible power series U such that

1) The linear part $B_i = \text{lin } G_1'(G_i' := U^{-1}G_iU)$ is in $\Delta_{n,\tau}$ for all $1 \leq i \leq m$.
2) All the G_i's are D_j-normal forms for all j's, where $D_j = d(B_j)$ is the diagonal part of B_j.

Proof. According to theorem 1 there is some linear transformation U_1, such that $U_1^{-1}.\text{lin } G_i.U_1$ is in $\Delta_{n,\tau}$ for some suitable τ. Thus we may suppose $\text{lin } G_i =: B_i \in \Delta_{n,\tau}$ and lin U = E. Now we prove the theorem by induction on m. The case m = 1 is a consequence of theorem 4 for $\vartheta = \{E\}$. Suppose secondly the theorem to be true for $G_1,\ldots,G_{m-1}$ and let $U'' = EX + \ldots$ be such that $G_i'' = U''^{-1}G_iU''$ commutes with $D_j = d(B_j)$ for $1 \leq i,j \leq m-1$. Then $G_m'' = U''^{-1}G_mU''$ commutes with all the G_i'', $1 \leq i \leq m-1$ implying G_m'' to be a D_j-normal form for $1 \leq j \leq m-1$ because of remark 3.

Using theorem 4 we get some $U' = EX + \sum_{1=2}^{\infty} u_1, U'$ D_j-normal form for $1 \leq j \leq m-1$, such that $G_m' := U'^{-1}G_m''U'$ is a ϑ-normal form for $\vartheta = \{D_1,\ldots,D_m\}$. Putting $U = U''U'$ we see that $G_j' := U''^{-1}G_j''U'' = U^{-1}G_jU$ is a ϑ-normal form (remark 3 is used again) for $1 \leq j \leq m-1$. Thus we get the desired result.

The following again is needed for the proof of the main theorem in this section.

Theorem 6. Let $F_1,\ldots,F_m,F_{m+1},\ldots,F_{m+1}$ be invertible formal power-series commuting in pairs, such that for some type τ the linear parts $B_i = \text{lin } F_i$ are in $\Delta_{n,\tau}$. Furthermore with $D_i = d(B_i)$ put $\vartheta = \vartheta' \cup \vartheta''$, $\vartheta' = \{D_1,\ldots,D_m\}$, $\vartheta'' = \{D_{m+1},\ldots,D_{m+1}\}$ and suppose $\mathcal{P}_i(\vartheta) = \mathcal{P}_i(\vartheta')$ for all $2 \leq i \leq k$. Then under the additional assumption, that all the F_j's are ϑ'-normal forms we have:

There exists some transformation $U = EX + \sum_{i>k} u_i$, U a $\mathcal{D}'$-normal form, such that $F_j' := U^{-1}F_jU$ is a $\mathcal{D}$-normal form for all $1 \leq j \leq m+1$.

Proof. The theorem can be proved by induction on ℓ, where Corollary 1 and theorem 4 are essential tools.

Lemma 2. Let $(T_m)_{m \in \mathbb{N}}$ be a sequence in Ω with $T_m = EX + \sum_{\ell \geq m} u_{m,\ell}$, $u_{m,\ell} \in \mathcal{P}_\ell$.

Then: i) The sequence $(S_m)_{m \in \mathbb{N}}$, $S_m = T_1T_2\ldots T_m = S_{m-1}T_m$ is convergent with respect to the topology induced by the order of formal power series. Exactly speaking we have, that $(S_m)_i$ equals $(S_{m+1})_i$ for $1 \leq i \leq m$.

ii) $S := \lim S_m$ is invertible and $S^{-1} = \lim S_m^{-1}$

iii) $(S^{-1} \circ F \circ S)_i = (S_j^{-1} \circ F \circ S_j)_i$ for all $j \geq i$.

Now we start to prove theorem 3

Proof (of theorem 3). According to theorem 1 we may suppose that $\text{lin}\, F \in \Delta_{n,\tau}$ for all $F \in \mathcal{F}$. Let $\mathcal{D}$ denote the set of diagonal parts $d(B_F)$ for $F \in \mathcal{F}$, where $B_F = \text{lin}\, F$. Since the vector spaces $\mathcal{P}_i = \mathcal{P}_i(\{E\})$ are of finite dimension, there exists a sequence $(\mathcal{D}_k)_{k \in \mathbb{N}, k \geq 2}$ of finite subsets of $\mathcal{D}$ such that $\mathcal{D}_k \subseteq \mathcal{D}_{k+1}$ for all k and such that $\mathcal{P}_i(\mathcal{D}) = \mathcal{P}_i(\mathcal{D}_k)$ for $2 \leq i \leq k$. Furthermore determine finite subsets $\mathcal{F}_k$ of $\mathcal{F}$ with $\mathcal{F}_k \subseteq \mathcal{F}_{k+1}$ and $\mathcal{D}_k = \{d(B_F) | F \in \mathcal{F}_k\}$. Taking $\mathcal{D}_1 := \phi$ we claim.

$$(*) \quad \begin{cases} \text{There exists a sequence } (T_k)_{k \geq 2}, T_k = EX + \sum_{j>k} u_{k,j}, u_{k,j} \in \mathcal{P}_j(\mathcal{D}_{k-1}) \text{ such that} \\ \text{with } S_k := T_2T_3\ldots T_k \text{ the power series } S_k^{-1}FS_k \text{ have to be } \mathcal{D}_k\text{-normal forms} \\ \text{for all } F \in \mathcal{F}_k. \end{cases}$$

To prove this by induction on k we see that theorem 4 gives us the case $k = 2$. Assuming now that $T_2, \ldots, T_k$ having these properties have been found we get T_{k+1} in the following way: Write $\mathcal{D}_{k+1} = \mathcal{D}_k \dot\cup \mathcal{D}'$. According to theorem 6 and using $\mathcal{P}_i(\mathcal{D}_k) = \mathcal{P}_i(\mathcal{D}) = \mathcal{P}_i(\mathcal{D}_{k+1})$ for $2 \leq i \leq k$ there exists $T_{k+1} = EX + \sum_{j>k+1} u_{k+1,j}$, $u_{k+1,j} \in \mathcal{P}_j(\mathcal{D}_k)$, such that $T_{k+1}^{-1}(S_k^{-1}FS_k)T_{k+1}$ is a $\mathcal{D}_{k+1}$-normal form for all $F \in \hat{\mathcal{F}}_{k+1}$.

Defining $S := \lim_{m\to\infty} S_m = S_k \circ \lim_{m\to\infty}(T_{k+1}T_{k+2}\ldots T_m) = S_kU_k$ we see that U_k is a $\mathcal{D}_k$-normal form because all $T_j, j \geq k+1$, are. Then it is easy to recognize that $F' := S^{-1}FS$ for $F \in \mathcal{F}_k$ is a $\mathcal{D}_k$-normalform. Taking $G \in \mathcal{F}$ arbitrary and using remark 3 we conclude that $G' := S^{-1}GF$ is a normal form with respect to all $D \in \bigcup_{k \in \mathbb{N}} \mathcal{D}_k$. This, finally, means by the very construction of the $\mathcal{D}_k$'s, that G' is a $\mathcal{D}$-normal-form.

§3. ITERATION PROBLEMS

It is well known (see, e.g.[1],[4],[8]) that a mapping $t \to F_t \in \Omega$ $(t \in \mathbb{Q},\mathbb{R},\mathbb{C})$ is called an iteration for F, if $F_{t+s} = F_t F_s$ for all t,s in the domain of definition and if $F_1 = F$. This iteration is called continuous, if all the mappings $t \to (F_t)_\nu$ are continuous. Iterations over $\mathbb{R}$ or $\mathbb{C}$ are called analytic, if the above mentioned mappings are analytic.

Reformulating a result of [1] we get after some calculations.

Theorem 7. Given $F \in \Omega$ invertible and given B with $\exp B = A = \text{lin } F$ there is an analytic iteration F_t for F with $\text{lin } F_t = \exp(tB)$ if and only if there is some transformation T fulfilling

1) $J := \text{lin}(T^{-1}BT)$ is a Jordan canonical form

2) $T^{-1}FT$ commutes with $d(\exp J)X$ for all t.

Additionally we have that for F commuting with all $d(\exp tJ)$ there is exactly one analytic iteration $F_t = (\exp tJ)X + \ldots$ such that F_t commutes with $d(\exp sJ)$ for all t,s. Furthermore the coefficients of $d(\exp(-tJ)).\ F_t$ are polynomials in t.

Definition 3. Given iterations $F_1(t),\ldots,F_m(t)$ of $F_1,\ldots,F_m$ respectively we say that these iterations commute in pairs, if $F_i(t)F_j(s) = F_j(s)F_i(t)$ holds for all t,s and all $1 \leq i,j \leq m$.

Then we get the following result generalizing a result on continuous iterations given in [5],[6].

Theorem 8. Let $F_1,\ldots,F_m \in \Omega$ commute in pairs. Then we have: There exist analytic iterations of $F_1,\ldots,F_m$ commuting in pairs if and only if there are analytic iterations $A_1(t),\ldots,A_m(t)$ of the linear parts $A_1,A_2,\ldots,A_m$ commuting in pairs and if there is some typ τ and some transformation T such that

1) $\text{lin}(T^{-1}A_i(t)T) =: B_i(t)$ is an element of $\Delta_{n,\tau}$ for all t's and all i's

2) $T^{-1}F_iT$ is a ϑ-normal form for all i's, where $\vartheta = \{d(B_j(t)) \mid 1 \leq j \leq m,\ t \in \mathbb{C}(\mathbb{R})\}$.

The proof is omitted.

The following theorem generalizes a theorem of D.C.Lewis [7].

Theorem 9. Given $F_1,\ldots,F_m$ invertible and commuting in pairs there are iterative powers $G_1 = F_1^{k_1},\ldots,G_m = F_m^{k_m}$ having analytic iterations commuting in pairs. Finally I want to mention a result on iterations without regularity conditions.

Theorem 10. Let $(F_t)_{t \in \mathbb{C}}$ be an iteration for F. Then there exists some transformation T such that

1) The linear part $\mathrm{lin}(T^{-1}F_tT) = B(t)$ is for all t an element of $\Delta_{n,\tau}$ for some τ.

2) $T^{-1}F_sT$ commutes with $d(B(t))$ for all s and t.

Proof. This is an immediate consequence of theorem 3.

Remark. Taking s = 1 in 2 we see that the existence of some T with

1) and

2) $T^{-1}FT$ commutes with $d(B(t))$ for all t

is necessary for the existence of an iteration for F with given linear part. This criterion is also sufficient as can be shown and is shown by G.Mehring in [13].

REFERENCES

[1] Reich, L., Schwaiger, J.: Über einen Satz von Shl.Sternberg in der Theorie der analytischen Iterationen. Monatsh.Math. 83, 207-221 (1977).

[2] Schwaiger, J., Reich, L.: Über die Funktionalgleichung N   T = T   M für formale Potenzreihen. Aequ. Math. 19, 66-78 (1979).

[3] Reich, L., Schwaiger, J.: Eine Linearisierungsmethode für Funktionalgleichungen vom iterativen Typus in Potenzreihenringen. Aequ. Math. 20, 224-243 (1980).

[4] Chen, K.T.: Local Diffeomorphismus - C^{∞} - Realizations of Formal Properties. Am. Journ. Math. 87, 140-157 (1965).

[5] Bucher, W.: Kontinuierliche Iterationen formal-biholomorpher Abbildungen. Ber.d.Math.-Stat.Sektion im Forschungszentrum Graz, Bericht Nr. 97 (1978).

[6] Reich, L.: Über kontinuierliche Iteration formal biholomorpher Abbildungen und Differenzenrechnung. Abh. Math. Sem. Hamburg 52 (1982).

[7] Lewis, D.C.: On Formal Power Series Transformations. Duke Math. J. 5, 794-805 (1939).

[8] Praagman, C.: Iteration and Logarithms of Formal Automorphisms, Preprint 1983.

[9] Jacobson, N.: Lectures in Abstract Algebra. Vol. II, Van Nostrand, Princeton, New Jersey, 1953.

[10] Greub, W.: Linear Algebra, 4th ed. Springer, 1975.

[11] Scheinberg, S.: Power Series in One Variable. Journ. Math. Anal. Appl. 31, 321-333.

[12] Reich, L., Kräuter, A.R.: Roots and Analytic Iteration of Formally Biholomorphic Mappings. Math.Rep.Canad.Acad.Sci. 3, 221-224 (1981).

[13] Mehring, G.: On a criterion of iteration in rings of formal power series. This volume.

Jens Schwaiger
Institut für Mathematik, Univ. Graz
Brandhofgasse 18, A-8010 Graz

ON INCREASING ITERATION SEMIGROUPS OF MULTI-VALUED FUNCTIONS

A. Smajdor

Let X be a non-empty set. A family $\{f^t : t \geq 0\}$ of multi-valued functions from X into X is called an iteration semigroup iff

1^0 $f^t(x)$ is a non-empty subset of X for every $x \in X$ and $t \geq 0$,

2^0 $f^s[f^t(x)] = f^{s+t}(x)$ for every $x \in X$ and $t,s \geq 0$, where

$f^s[f^t(x)] = \cup \{f^s(y) : y \in f^t(x)\}$ is the image of the set $f^t(x)$ under f^s (see [1] Chapter II).

An iteration semigroup $\{f^t : t \geq 0\}$ is said to be increasing iff the inclusion $f^t(x) \subset f^s(x)$ holds for every $x \in X$ and $s \geq t \geq 0$.

An iteration semigroup $\{f^t : t \geq 0\}$ such that $f^1 = f$ is called an iteration semigroup of f.

The following theorem will be used in the paper.

Theorem 1 (see [3]). Let X be a non-empty set and let $\{f^t : t \geq 0\}$ be an increasing iteration semigroup such that

(H_1) $\bigcup_{t \geqq 0} f^t(x) = X \quad (x \in X)$,

(H_2) the sets $\{t \geq 0 : y \in f^t(x)\}$ $(x,y \in X)$ are closed,

and

(H_3) $y \in f^t(x)$ if and only if $x \in f^t(y)$ for every $x,y \in X$, $t \geq 0$. Then the function

$$a(x,y) : = \inf \{t \geq 0 : y \in f^t(x)\}$$

fulfills the following conditions

(a) a is finite,

(b) for every $t \geq 0$ and $x \in X$ there exists $y \in X$ such that $a(x,y) \leq t$,

(c) $a(x,z) \leq a(x,y) + a(y,z)$ for every $x,y,z \in X$,

(d) for every $t,s \geq 0$, $x,z \in X$ with $a(x,z) \leq t + s$ there exists $y \in X$ such that $a(x,y) \leq t$, $a(y,z) \leq s$,

(e) $f^t(x) = \{y \in X : a(x,y) \leq t\}$ for $t \geq 0$ and $x \in X$,

and

(f) $a(x,y) = a(y,x)$ for every $x,y \in X_0$.

We define functions

$$p_{\bar{x}}(x) = a(x,\bar{x})$$

for $\bar{x},x \in X$.

Now, we shall prove the following

Theorem 2. Let X be a non-empty set and let $\{f^t : t \geq 0\}$ be an increasing iteration semigroup fulfilling conditions (H_1) - (H_3). Then

$$f^t(x) = \bigcap_{\bar{x} \in X} p_{\bar{x}}^{-1}[p_{\bar{x}}(x) + tB] ,$$

where $B = [-1,1]$.

Proof. According to conditions (c) and (f) we have

$$|p_{\bar{x}}(x) - p_{\bar{x}}(y)| \leq a(x,y)$$

for $\bar{x},x,y \in X$. It implies

$$\sup_{\bar{x} \in X} |p_{\bar{x}}(x) - p_{\bar{x}}(y)| \leq a(x,y) \tag{1}$$

for every $x,y \in X$.

We put

$$a(y,y) = 3t . \tag{2}$$

In virtue of (d) there exists $z \in X$ such that

$$a(y,z) \leq t \quad \text{and} \quad a(z,y) \leq 2t . \tag{3}$$

By (2), (3), (c) and (f) we have

$$3t = a(y,y) \leq a(y,z) + a(z,y) = 2a(y,z) \leqq 2t .$$

Consequently $t = 0$. Thus $p_y(y) = a(y,y) = 0$.

Now we have

$$a(x,y) = |p_y(x) - p_y(y)| = \sup_{\bar{x} \in X} | p_{\bar{x}}(x) - p_{\bar{x}}(y)| . \tag{4}$$

Inequalities (1) and (4) give

$$a(x,y) = \sup_{\bar{x} \in X} |p_{\bar{x}}(x) - p_{\bar{x}}(y)|$$

for $x,y \in X$.

The inequality

$$\sup_{\bar{x} \in X} |p_{\bar{x}}(x) - p_{\bar{x}}(y)| \leq t$$

is equivalent to

$$p_{\bar{x}}(y) \in p_{\bar{x}}(x) + tB$$

for $\bar{x} \in X$, where $B = [-1,1]$. Therefore condition (e) can be rewritten in the form

$$f^t(x) = \bigcap_{\bar{x} \in X} p_{\bar{x}}^{-1}[p_{\bar{x}}(x) + tB] .$$

This completes the proof.

According to (d), (a) and (1) we obtain

Corollary 1. If conditions (H_1) - (H_3) hold, then

(C_1') if $x,z \in X$, $|p_{\bar{x}}(x) - p_{\bar{x}}(z)| \leq t + s$ for every $\bar{x} \in X$, then there exists $y \in X$ such that $|p_{\bar{x}}(x) - p_{\bar{x}}(y)| \leq t$, $|p_{\bar{x}}(y) - p_{\bar{x}}(z)| \leq s$ for every $\bar{x} \in X$, and

(C_2') for every $x,y \in X$ there exists $t \geq 0$ such that $|p_{\bar{x}}(y) - p_{\bar{x}}(x)| \leq t$ for every $\bar{x} \in X$.

If X is a metric space and $\{f^t : t \geq 0\}$ is an iteration semigroup of compact-valued functions from X into X such that the multi-valued function

$$(t,x) \to f^t(x) \tag{5}$$

is closed, then the function a is lower semicontinuous (see [3]). Thus we have

Corollary 2. We suppose that X is a metric space and $\{f^t : t \geq 0\}$is an increasing iteration semigroup of compact-valued functions from X into X for which (H_1) - (H_3) are fulfilled and the multi-valued function (5) is closed. Then there exists a family P of real lower semicontinuous functions such that

$$f^t(x) = \bigcap_{p \in P} p^{-1}[p(x) + tB]$$

for $x \in X$, $t \geq 0$, where $B = [-1,1]$ and conditions

(C_1) if $|p(x) - p(z)| \leq t + s$ for every $p \in P$ then there exists $y \in X$ such that $|p(x) - p(y)| \leq t$ and $|p(y) - p(z)| \leq s$ for every $p \in P$,

(C_2) for every $x,y \in X$ there exists $t \geq 0$ such that $|p(y) - p(x)| \leq t$ for every $p \in P$,

are fulfilled.

For the proof it suffices to put $P := \{p_{\bar{x}} : \bar{x} \in X\}$. The following theorem is the main result of this paper.

Theorem 3. Let X be a non-empty set and let f be a multi-valued function from X into X with non-empty values. Function f has an increasing iteration semigroup $\{f^t : t \geq 0\}$ fulfilling conditions (H_1) - (H_3) if and only if there exists a family P of real functions fulfilling conditions (C_1), (C_2) and such that

$$f(x) = \bigcap_{p \in P} p^{-1}[p(x) + B], \tag{6}$$

where $B = [-1,1]$.

Proof. The necessity is an immediate consequence of Theorem 2 and Corollary 1. To prove the sufficiency suppose that P is a given family of real functions fulfilling (C_1), (C_2) and (6). Let

$$f^t(x) = \bigcap_{p \in P} p^{-1}[p(x) + tB]$$

for $x \in X$, $t \geq 0$. It is obvious that $f^1(x) = f(x)$, $x \in f^t(x)$ and $f^t(x) \subset f^s(x)$ for $s \geq t \geq 0$, $x \in X$.

We suppose that $t,s \geq 0$ and $z \in f^{t+s}(x)$. It means that $|p(z) - p(x)| \leq t + s$. By (C_1) there exists $y \in X$ such that $|p(y) - p(x)| \leq t$ and $|p(z) - p(y)| \leq s$ for every $p \in P$. Thus $y \in f^t(x)$, $z \in f^s(y)$, so $z \in f^s[f^t(x)]$. The inclusion

$$f^{s+t}(x) \subset f^s[f^t(x)] \tag{7}$$

is proved.

Now we suppose that $z \in f^s[f^t(x)]$. Then there exists $y \in f^t(x)$ such that $z \in f^s(y)$. Therefore $|p(y) - p(x)| \leq t$ and $|p(z) - p(y)| \leq s$ for every $p \in P$. Hence $|p(z) - p(x)| \leq |p(z) - p(y)| + |p(y) - p(x)| \leq s + t$ for every $p \in P$ and thus $z \in f^{s+t}(x)$. This proves the inclusion

$$f^s[f^t(x)] \subset f^{s+t}(x) \quad , \tag{8}$$

for $x \in X$ and $t,s \geq 0$. By (7) and (8) the family $\{f^t : t \geq 0\}$ is an iteration semigroup.

Applying (C_2) for every $x,y \in X$, there exists $t \geq 0$ such that $|p(y) - p(x)| \leq t$, $p \in P$. Therefore $y \in f^t(x)$ for certain $t \geq 0$ and condition (H_1) is fulfilled. Since the equality

$$\{t \geq 0 : y \in f^t(x)\} = \bigcap_{p \in P} \{t \geq 0 : |p(y) - p(x)| \leq t\}$$

holds for $x,y \in X$ the set $\{t \geq 0 : y \in f^t(x)$ is closed. Condition (H_2) is fulfilled. Condition (H_3) is obvious. This completes the proof.

We shall consider a particular case. Let $X = (0,a)$ or $[0,a)$, $0 < a \leq +\infty$ and let $g : X \to X$ be a continuous and strictly increasing function such that $0 < g(x) < x$ in X. Fix $x_0 \in X$ and write $X_0 = [g(x_0),x_0]$. Every continuous and decreasing function $P_0 : X_0 \to R$ fulfilling the condition

$$p_0[g(x_0)] = p_0(x_0) + 1$$

can be uniquely extended onto X to a continuous and decreasing solution $p : X \to R$ of the Abel functional equation

$$p[g(x)] = p(x) + 1 \tag{9}$$

(see [2], Theorem 2.1). We suppose that $\lim_{x \to a-} g(x) = a$. Then the function g^{-1} is defined in X and

$$p[g^{-1}(x)] = p(x) - 1 \tag{10}$$

for $x \in X$. Let

$$f(x) = [g(x), g^{-1}(x)]$$

for $x \in X$. By (9), (10), continuity and monotonicity of p we have

$$p[f(x)] = [\, p[g^{-1}(x)],\ p[g(x)]] = p(x) + [-1,1].$$

We shall show that conditions (C_1) and (C_2) are satisfied. Suppose that $|p(x)-p(z)|=t + s$, where $x,z \in X$ and $t,s \geq 0$. If $t = 0$, then $|p(x) - p(y)| = t$, $|p(y) - p(z)| = s$ for $y = x$. Consider the case $t > 0$, $s > 0$, $x < z$. The function $q(y) = |p(y) - p(x)|$ is continuous and $q(x) = 0$ and $q(z) = p(x) - p(z) = t + s > t$. Therefore there exists $y \in X$, $x < y < z$ such that $q(y) = p(x) - p(y) = t$. Hence $|p(z) - p(y)| = p(y) - p(z) = p(y) - p(x) + p(x) - p(z) = - t + t + s = s$. In the case $z < x$ the proof of (C_1) is similar. Since the family $P = \{p\}$ has only one element it fulfils (C_2). According to Theorem 3 the function f has an increasing iteration semigroup fulfilling conditions (H_1) - (H_3).

REFERENCES

[1] Berge, C.: Topological spaces, Oliver & Boyd, Edinburgh and London 1963.

[2] Kuczma, M.: Functional equations in a single variable, Monografie Mat. 46 PWN, Warzawa 1968.

[3] Smajdor, A.: Iterations of multi-valued functions, in preparation.

A. Smajdor
Instytut Matematyki
Uniwersytetu Śląskiego
ul. Bankowa 14
PL-40-007 Katowice

PLANT GROWTH AS AN ITERATION PROCESS

W. Szlenk, W. Żelawski*)

ABSTRACT. A model of vegetative plant growth is formulated on the base of experimental data of dry matter accumulation and distribution in seedlings of Scots pine. The quantitative description of growth is carried out in terms of discrete dynamical systems. Main results: if time tends to infinity, then: (1) the proportion of dry weight of assimilatory to nonassimilatory parts tends to a fixed value; (2) accumulation of dry weight approaches a function Pt + Q + R ln t (t-time, P,Q,R-constants).
AMS subject classifications: 58F Dynamical systems 92A09 Physiology. Key words: discrete dynamical system, stationary point, vegetative plant growth, partitioning of dry matter.

0. There is an increasing interest towards the quantitative approach in plant growth and productivity studies (see e.g.[1,2,3,4]). However, some popular growth functions, which are being used in growth analysis of either animals or whole populations, are of very little help in growth analysis of plants due to significant differences in life strategies of various organisms. Whereas animals gain their weight through consumption of food that is taken up in a form of organic matter, the plant body is formed through an autotrophic process occuring in photosynthesizing leaves. Thus, although the growth of a plant is a sum of all its growing parts, the role played in the total productivity by assimilatory organs themselves is a special one; the partitioning of photosynthethic products into assimilatory and nonassimilatory parts which varies phenotypically is of great importance for growth of the whole plant. Just such a compartmentization appeared to be a reasonable simplification at modeling plant growth and so it had been observed throughout the whole text of the paper.

In this article we outline only the main results. The full text has been submitted to Acta Aplicandae Mathematicae; see also [7].

1. There are two main features dependent on availability of such extremal factors as water, fertilizers, light, carbon dioxide, etc. There are - the efficience of assimilatory organs (so called unit leaf rate or net assimilation rate [2]) and the partitioning of photosynthetic products between the two, above mentioned compartments. At first approach the two parameters, characterizing both features could be considered as the constant values with respect to the influence of external conditions, but in reality they may vary not only because of the already accomplished

---

*)Presented by W. Szlenk

growth but also due to weather fluctuations and other changes in the environment. This should be taken into account at more detailed approach.

All experimental data the model is based on, come from small seedlings of Scots pine (Pinus silvestris L) grown during their first growing season; they were the object of research for many years in the plant Physiology Laboratory of the Warsaw Agricultural University (for details see [5,6]). The simple simulation experiments were carried out using the pocket calculator HP-33C whereas more complicated ones by use of the computer RIAD at the Computer Center of this University.

In mathematical terms the growth model is a discrete two dimensional dynamical system, with usually one day as time unit. It is described in the second section of the paper, also some asymptotic properties of the system are proved. Specification of parameters and a theorem giving the asymptotic growth function comes in the third section. In the fourth section we further specify some parameters and we present a comparison of the theoretical curves and experimental data.

2. Let W_n and V_n denote, respectively, biomass (dry weight) of assimilatory and non assimilatory parts respectively at the n-th unit of time. Set

$$M_n = W_n + V_n$$

$$\lambda_n = \frac{W_n}{V_n} \quad . \tag{1}$$

The number M_n is the total dry weight of the plant, and λ_n is the proportion of its assimilatory and non assimilatory parts. Obviously $M_n \geq 0$, $\lambda_n \geq 0$.

We assume that the dynamics of the growth process in time is given by the equations:

$$W_{n+1} = W_n + \beta(\lambda_n,\delta)\alpha(W_n)W_n$$

$$V_{n+1} = V_n + 1 - \beta(\lambda_n,\delta)\ \alpha(W_n)W_n \ , \qquad n = 0,1,2,... \tag{2}$$

where $\alpha(W)$ is a decreasing, continuous function of W, $\alpha(0) = \alpha_0 > 0$, $\lim_{W\to+\infty} \alpha(W) = 0$; the function $\beta(\lambda,\delta)$ is a decreasing, continuous function with respect to λ, $\beta(0,\delta) = 1$ and $\lim_{\lambda\to+\infty} \beta(\lambda,\delta) = 0$. The coefficient δ, $0 \leq \delta \leq 1$, represents an action of enviromental conditions upon the proportion λ. At the first approach δ is assumed to be constant. That is why we shall write $\beta(\lambda)$ instead of $\beta(\lambda,\delta)$. The coefficient $\alpha(W)$ expresses the unit leaf rate ([2]) i.e. $\alpha(W_n)W_n$ is the amount of dry substance produced by the whole assimilatory part W_n in the n-th unit of time. If the total dry weight of the assimilatory part increases (then, of course the total mass M also increases) the coefficient $\alpha(W)$ declines due to selfshading effects and maintenance costs of the whole system.

The number $\beta(\lambda_n)$ represents the partitioning coefficient which says how much of the newly produced material $\alpha(W_n)W_n$ is being used to enlarge the assimilatory part itself. If $\lambda_n \approx 0$, i.e. $V_n >> W_n$, then $\beta(\lambda_n) \approx 1$ which means that almost all material $\alpha(W_n)W_n$ is used to build the assimilatory part. If λ_n is large, i.e. $V_n << W_n$, then $\beta(\lambda_n) \approx 0$ which means that almost all material $\alpha(W_n)W_n$ is used for extension of the non-assimilatory part.

The process of growth described by the equations (2), n = 0,1,..., is simply the family of the iterates of the following mapping:

$$W_1 = W + \beta(\lambda)\alpha(W)W$$

$$V_1 = V + [1 - \beta(\lambda)]\alpha(W)W \ ,$$

$\lambda = \frac{W}{V}$, acting on the domain $\{W > 0, V > 0\}$.

It is useful to consider the mapping above in the coordinates (M,λ): the formula (1) gives us

$$V = \frac{1}{1+\lambda} M, \quad W = \frac{\lambda}{1+\lambda} M \quad .$$

Therefore the mapping in question takes the following form:

$$M_1 = M + \alpha(\frac{\lambda}{1+\lambda} M) \frac{\lambda}{1+\lambda} M =: M_1(M,\lambda)$$

$$\lambda_1 = \frac{W_1}{V_1} = \frac{1+\beta(\lambda)\alpha(\frac{\lambda}{1+\lambda} M)}{1+[1-\beta(\lambda)]\alpha(\frac{\lambda}{1+\lambda} M)\lambda} \lambda =: \lambda_1(M,\lambda) \quad . \tag{3}$$

Setting

$$f(M,\lambda) = \frac{1 + \beta(\lambda)\alpha(\frac{\lambda}{1+\lambda} M)}{1 + [1-\beta(\alpha)]\alpha(\frac{\lambda}{1+\lambda} M)\lambda} \tag{4}$$

the second equation in (3) takes the form

$$\lambda_1(M,\lambda) = \lambda f(M,\lambda) \quad . \tag{5}$$

The configuration space for the system given by (3) is the domain $D = \{(M,\lambda): M \geq 0, \lambda \geq 0\}$. The equations (3) define a mapping ϕ of D into itself: for a point $(M,\lambda) \in D$ we have

$$\phi(M,\lambda) = (M_1(M,\lambda), \lambda_1(M,\lambda)) \in D \ ;$$

the symbol ϕ^n denotes the n-th iterate of ϕ, i.e. $\phi^n = \phi \circ \ldots \circ \phi$, therefore after n units of time we have

$$(M_n,\lambda_n) = \phi(M_{n-1},\lambda_{n-1}) = \phi^n(M,\lambda).$$

The growth process is simply a discrete dynamical system (D,ϕ), with the configuration space D being a two dimensional domain in the plane.

Proposition 1. The mapping ϕ is an endomorphism of D into, but not onto itself.

We omit the proof.

Denote by $\bar{\lambda}$ a number such that

$$\bar{\lambda} = \frac{\beta(\bar{\lambda})}{1-\beta(\bar{\lambda})} \quad . \tag{6}$$

Since $\beta(\lambda)$ is a decreasing function and $\beta(0) = 1$, $\beta(+\infty) = 0$, such a point $\bar{\lambda} > 0$ does exist and is unique.

We shall assume from now on that $\alpha(W)$ and $\beta(\lambda)$ are of class C^1.

Lemma 1. There exists a number $\bar{M} > 0$ such that if $M > \bar{M}$ and $0 \leq \lambda \leq \bar{\lambda}$ then

$$0 \leq \lambda_1(M,\lambda) \leq \bar{\lambda} \quad .$$

For the proof see [7].

Proposition 2. For any $M_o > 0$, $\lambda_o > 0$ holds

$$\lim_{n\to+\infty} M_n = +\infty .$$

Proof. In view of (3) we get by induction

$$M_n = M_0 \prod_{i=0}^{n-1} \left[1 + \frac{\lambda_i}{1+\lambda_i} \alpha\left(\frac{i}{1+\lambda_i} M_i\right)\right] \tag{7}$$

where M_0 is the initial dry weight at the beginning of growth. Obviously the sequence (M_n) is increasing. Suppose it is bounded, i.e. that there exists a number N such that for any n

$$M_n \leq N \quad .$$

The function $\lambda_1(M,\lambda)$ is bounded in the domain $\{0 \leq M \leq N, \lambda > 0\}$. Since $\lambda_1(M,\lambda) \geq \lambda$

for $0 \leq \lambda \leq \bar{\lambda}$ (see below), we conclude that there exists a number $\delta > 0$ such that $\lambda_i \geq \delta$ for each $i = 0,1,\ldots$.

Therefore

$$\frac{\lambda_i}{1+\lambda_i}\,\alpha\left(\frac{\lambda_i}{1+\lambda_i}\,M_i\right) \geq \frac{\delta}{1+\delta}\,\alpha\left(\frac{\delta}{1+\delta}\,N\right) =:\theta \quad .$$

Hence, in view of (10), we have

$$M_n \geq M_o \prod_{i=0}^{n-1} (1+\theta) = M_o(1+\theta)^n \qquad n = 0,1,\ldots$$

which contradicts $M_n \leq N$.

Corollary 1. The sequence $(M_n)_o^\infty$ does not tend to infinity exponentially, i.e. there do not exist numbers $\rho > 1$ and $A > 0$ such that $M_n \geq A\,\rho^n, n = 0,1,\ldots$; it follows easily from the formula (10).

Theorem 1. Assume $\alpha(W)$ and $\beta(\lambda)$ to be of class C^1 and satisfy the conditions formulated on page 3. Then for any $M_o > 0$, $\lambda_o > 0$ holds

$$\lim_{n\to+\infty} \lambda_1(M_n,\lambda_n) = \bar{\lambda} \quad .$$

Proof. The function $\lambda_1(M,\lambda)$ has exactly one positive fixed point $\lambda = \bar{\lambda}(\lambda = 0)$ is also a fixed point). Indeed, if

$$\lambda_1(M,\lambda) = \frac{1 + \beta(\lambda)\alpha(\frac{\lambda}{1+\lambda}\,M)}{1+[1-\beta(\lambda)]\alpha(\frac{\lambda}{1+\lambda}\,M)\lambda}\,\lambda = \lambda$$

then

$$1 + \beta(\lambda)\alpha\left(\frac{\lambda}{1+\lambda}\,M\right) = 1+[1-\beta(\lambda)]\alpha\left(\frac{\lambda}{1+\lambda}\,M\right)\lambda$$

which implies

$$\lambda = \frac{\beta(\lambda)}{1-\beta(\lambda)} \; ,$$

i.e. $\lambda = \bar{\lambda}$. We see also that for $0 \leq \lambda \leq \bar{\lambda}$

$$\lim_{\lambda\to 0} \frac{\lambda_1(M,\lambda)}{\lambda} = \lim_{\lambda\to 0} f(M,\lambda) = 1 + \alpha_o > 1 \; ,$$

which means that $\lambda_1(M,\lambda) \geq \lambda$ for $0 \leq \lambda \leq \bar{\lambda}$. Since $\lim_{\lambda\to+\infty} \lambda_1(M,\lambda) = \frac{1}{\alpha(M)}$, we conclude that $\lambda_1(M,\lambda) \leq \lambda$ for $\lambda \leq \bar{\lambda}$. Suppose that $M_o > 0$, $\lambda_o > 0$. In virtue of Proposition 2 there exists an integer n_0 such that for $n \geq n_0$, $M_n \geq \bar{M}$ where $\bar{M}$ is from Lemma 1. If $\lambda_{n_0} \in [0,\bar{\lambda}]$, then in view of Lemma 1

$$\lambda_{n_0} < \lambda_{n_0+1} < \bar{\lambda} \quad ,$$

and by induction

$$\lambda_{n_0} < \lambda_n < \lambda_{n+1} < \bar{\lambda} \quad , \tag{8}$$

for any $n \geq n_0$. If $\lambda_{n_0} \in [\bar{\lambda},+\infty)$, then either $\lambda_n \in [\bar{\lambda},+\infty)$ for all $n \geq n_0$, or there exists a λ_{n_1} such that $\lambda_{n_1} \in \langle 0,\bar{\lambda}]$ and then (11) holds for $n \geq n_1$. If $\lambda_n \in [\bar{\lambda},+\infty)$ for all $n \geq n_0$, then

$$\lambda_n > \lambda_{n+1} > \bar{\lambda} \tag{9}$$

for all $n > n_0$. The formulas (8) and (9) imply that there exists a limit

$$\lim_{n\to+\infty} \lambda_n =: g \ .$$

Suppose $g \neq \bar{\lambda}$. Without loss of generality we may assume $\lambda_{n_0} < \bar{\lambda}$ (in the other case the proof is similar). The sequence $(\lambda_n)_{n_0}^{\infty}$ is increasing, so

$$\lambda_{n+1} - \lambda_n = f(M_n,\lambda_n)\lambda_n - \lambda_n = \lambda_n[f(M_n,\lambda_n) - 1] = \tag{10}$$

$$= \lambda_n \left| \frac{1 - \beta(\lambda_n)\alpha(\frac{\lambda_n}{1+\lambda_n} M_n)}{1+[1-\beta(\lambda_n)]\alpha(\frac{\lambda_n}{1+\lambda_n} M_n)\lambda_n} - 1 \right| =$$

$$= \lambda_n \frac{1+\beta(\lambda_n)\alpha(\frac{\lambda_n}{1+\lambda_n} M_n)-1-[1-\beta(\lambda_n)]\alpha(\frac{\lambda_n}{1+\lambda_n} M_n)\lambda_n}{1+[1-\beta(\lambda_n)]\alpha(\frac{\lambda_n}{1+\lambda_n} M_n)\lambda_n} =$$

$$= \lambda_n \ (\frac{\lambda_n}{1+\lambda_n} M_n)[1-\beta(\lambda_n)] \frac{\frac{\beta(\lambda_n)}{1-\beta(\lambda_n)} - \lambda_n}{1+[1-\beta(\lambda_n)]\alpha(\frac{\lambda_n}{1+\lambda_n} M_n)\lambda_n} \geq$$

$$\geq \lambda_n \alpha(\frac{\lambda_n}{1+\lambda_n} M_n)[1 - \beta(\lambda_n)] \left| \frac{\beta(\lambda)_n}{1-\beta(\lambda_n)} - \lambda_n \right| .$$

The sequence $(\lambda_n)_{n_0}$ tends to $g > 0$, so that there exists a number $\delta > 0$ such that $\lambda_n \geq \delta$. Thus

$$\delta \leq \lambda_n \leq g < \bar{\lambda} \quad .$$

Applying this inequality to (10) and keeping in mind that $\beta(\lambda)$ is decreasing we get

$$\lambda_{n+1} - \lambda_n \geq \delta[1-\beta(\delta)] \left| \frac{\beta(g)}{1-\beta(g)} - g \right| \alpha(\frac{\lambda_n}{1+\lambda_n} M_n) = g_1 \alpha(\frac{\lambda_n}{1+\lambda_n} M_n) \tag{11}$$

where

$$g_1 = \delta[1-\beta(\delta)] \left| \frac{\beta(g)}{1-\beta(g)} - g \right| .$$

Summing (14) over all $n \geq n_0$ we get

$$g - \lambda_{n_0} \geq g_1 \sum_{n=n_0}^{\infty} \alpha(\frac{\lambda_n}{1+\lambda_n} M_n) \quad .$$

Thus in view of (7) we conclude that the product

$$\prod_{i=0}^{\infty} \left| 1 + \frac{\lambda_i}{1+\lambda_i} \alpha(\frac{\lambda_i}{1+\lambda_i} M_i) \right|$$

is convergent, which means that the sequence $(M_n)_0^\infty$ would be bounded. We have a contradiction with the statement of Proposition 2. So that $g = \bar{\lambda}$ i.e.

$$\lim_{n\to\infty} \lambda_1 (M_n, \lambda_n) = \bar{\lambda} \quad .$$

This completes the proof.

Remark 1. We have proved that if M_0 is large enough then the sequence (λ_n) is monotone. The theorem does not say anything about the behaviour of the sequence if M_0 is rather small. Actually one can expect that if λ_0 were close to 0, i.e. the plant was deprived of almost all green parts, and M_0 was also small, then the corresponding sequence would oscillate about $\bar{\lambda}$ (damped oscillations) at the beginning and afterward would become monotone.

Remark 2. The two extremal cases $\lambda = 0$ and $\lambda = +\infty$ are also biologically interpretable. The first one would denote a situation of the plant devoid of assimilatory organs (e.g. a deciduous tree in spring); the second one could be a shoot alone without root (e.g. a leaf or a stem before rooting when vegetatively propagated). For any M the points $\lambda = 0$ and $\lambda = +\infty$ are repulsive fixed points. It means that the plant tends to escape from the state where either assimilatory or non-assimilatory organs are lacking or are very much out of balance. Regeneration of the missing part is then very vigorous.

Remark 3. The statement of Theorem 1 is not quite valid in the real world. Although the plant may theoretically grow to infinity, its real growth is usually interrupted in a rather catastrophic way (wind, deterioration of the water regime, disease etc.). Under certain external conditions for W_n large enough $\alpha(W_n)$ would eventually approaches zero causing a complete cessation of growth. However, under natural conditions α happens to reach zero more quickly due to sudden deterioration of photosynthetic activity (for instance at the end of the growing season). Hence the plant growth curve is usually of sigmoidal shape.

Remark 4. The value of the stationary point $\bar{\lambda}$ is experimentally measurable. Growing the plant under constant conditions for some time one can achieve stabilization of the proportion of assimilatory and nonassimilatory organs (see Fig. 1).

3. In this section we consider a special form of the function $\alpha(W)$. On the basis of experimental data one can assume $\alpha(W)$ is of the form

$$\alpha(W) = \frac{a_0}{a+W} \tag{12}$$

which means that the reciprocal function

$$y = y(W) = \frac{1}{\alpha(W)} = \frac{1}{a_0} W + \frac{a}{a_0}$$

is linear. On Fig. 2 we present some experimental data.

Remark 5. The quantity of material produced in the n-th unit of time is equal to $\alpha(W_n)W_n$. If $\alpha(W)$ is of the form (15), the total production is equal to

$$\alpha(W_n)W_n = \frac{a_0}{a+W_n} W_n \quad .$$

So the rate of biomass production is proportional to the fraction $\frac{W_n}{a+W_n}$; this resembles the Michaelis-Menten equation usually applied for expressing the dependence of the rate of biochemical reaction upton the concentration of substrate.

Theorem 2. If (W) is of the form (12) and $\beta(\lambda)$ is of class C^1, then there exist three numbers P,Q,R such that

$$\lim_{n\to+\infty} [M_n - (P_n + Q \ln n + R)] = 0 \quad .$$

The proof is very technical and we do not present it here. The reader may find in [7].

Remark 6. The meaning of the Theorem 2 is as follows: for large n the total mass of the plant increases almost as a linear function (Pn + R) minus a term proportional to the logarithm n(Q ln n).The absolute error of the approximation $M_n \approx Pn + Q \ln n + R$ tends to 0 as n tends to infinity and is practically without significance.

Remark 7. It would be very difficult to check experimentally the result formulated in Theorem 2 because infinite growth does not occur in reality. If one were even able to keep the plant alive for a sufficiently long time, the technical difficulties of measurement would become invicible as it is the case e.g. with large trees. Hence, Theorem 2 and its proof have theoretic meaning only. Nevertheless if we consider growth curves of wood accumulation in trunks of large trees, we see that being at the beginning exponential they become gradually more and more similar to linear ones. This is exactly what Theorem 2 says: for n large enough the total mass M_n is asymptotically equal to Pn; the term Q ln n being negligible since $\frac{\ln n}{n} \to 0$.

4. Now we shall specify the function $\beta(\lambda) = \beta(\lambda,\delta)$. As yet we assumed only that $\beta(\lambda)$ is of class C^1, decreasing $\beta(0) = 1$, $\lim_{\lambda\to+\infty} \beta(\lambda) = 0$. From among various possibilities we postulate the following simple form of the function $\beta(\lambda)$:

$$\beta(\lambda) = \frac{\delta}{\delta+(1+\delta)\lambda} , \qquad 0 < \delta < 1 , \tag{13}$$

which fulfils the above conditions. The parameter δ expresses the environmental conditions (for instance watering, fertilization, etc.). The form (13) of $\beta(\lambda)$ gives simple partitioning coefficients in (2). Namely, taking into account that $\lambda_n = \frac{W_n}{V_n}$ we have

$$\beta(\lambda_n,\delta) = \frac{\delta V_n}{\delta V_n+(1-\delta)W_n} , \qquad 1-\beta(\lambda_n,\delta) = \frac{(1-\delta)W_n}{\delta V_n+(1-\delta)W_n} \quad .$$

The form (13) appears to be "empirically justified", e.g. the theoretical growth sequences (M_n) and the experimental ones are close to each to other (Fig. 3). The form (13) of $\beta(\lambda)$ gives a simple formula for the stationary value $\bar{\lambda}$. Namely, combining formulas (6) and (13) we get

$$\bar{\lambda} = \frac{\delta}{\sqrt{1-\delta}} .$$

Since $\bar{\lambda}$ is a proportion between assimilatory and nonassimilatory organs, experimentally measurable, δ can be also considered as a known quantity. In Fig. 3 one can see the theoretical growth curves as verified by experimental data $(M_n^{exp})_{n=1}^{5}$. The theoretical values M_n depend on two parameters: a_o and a (called in the Fig. 3 α and $W_{\alpha/2}$). For each experimental set of data these parameters were chosen in such a way that the sum of squares of errors of all points was minimalized. Since we have two parameters and five experimental points in one experiment and four in another one, the good fit of theoretical curves and experimental data speaks in favour of the model.

FIGURES

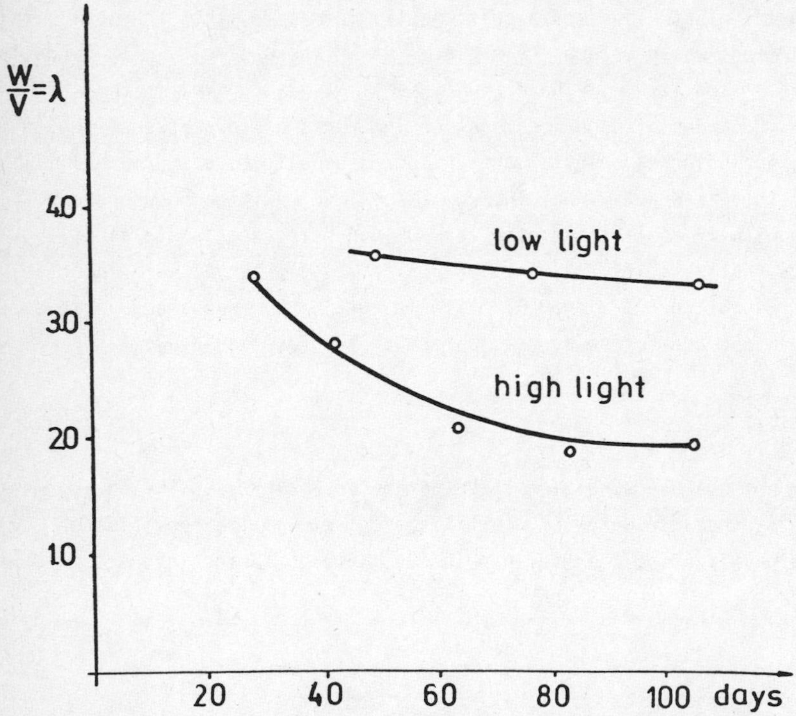

Fig. 1. Experimental data illustrating a trend in proportion of assimilatory to nonassimilatory parts λ in seedlings of Scots pine (Pinus silvestris L) grown in water culture under laboratory conditions at two different light intensities.

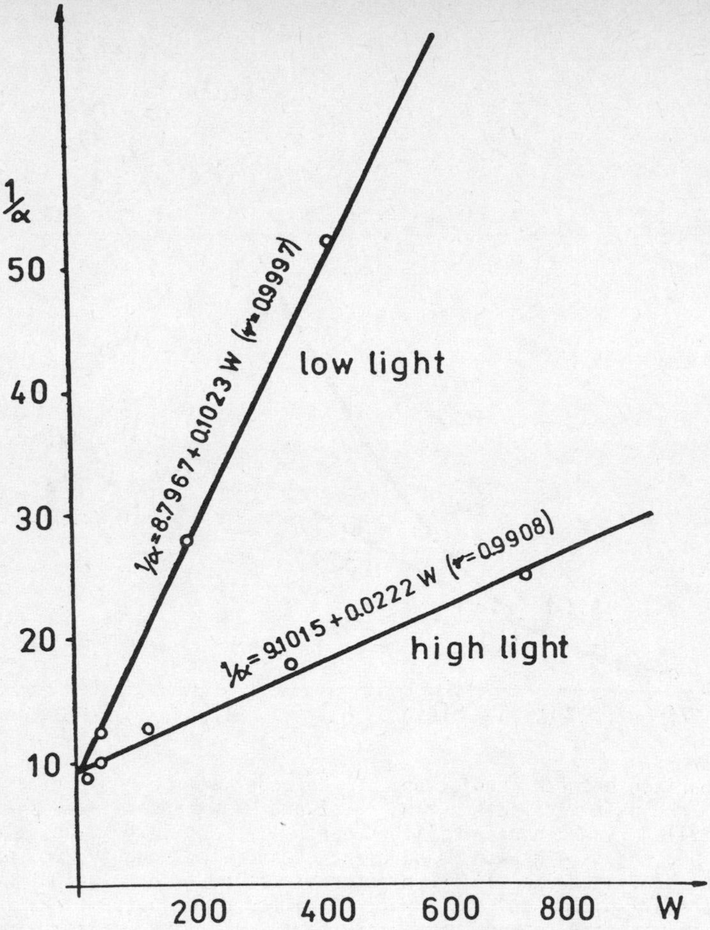

Fig. 2. Linear relationship between reciprocal of unit leaf rate and size of assimilatory organs $W = \lambda/(1+\lambda)M$ fitted, by least square method, to the data obtained from the same experiments as in Fig. 1.

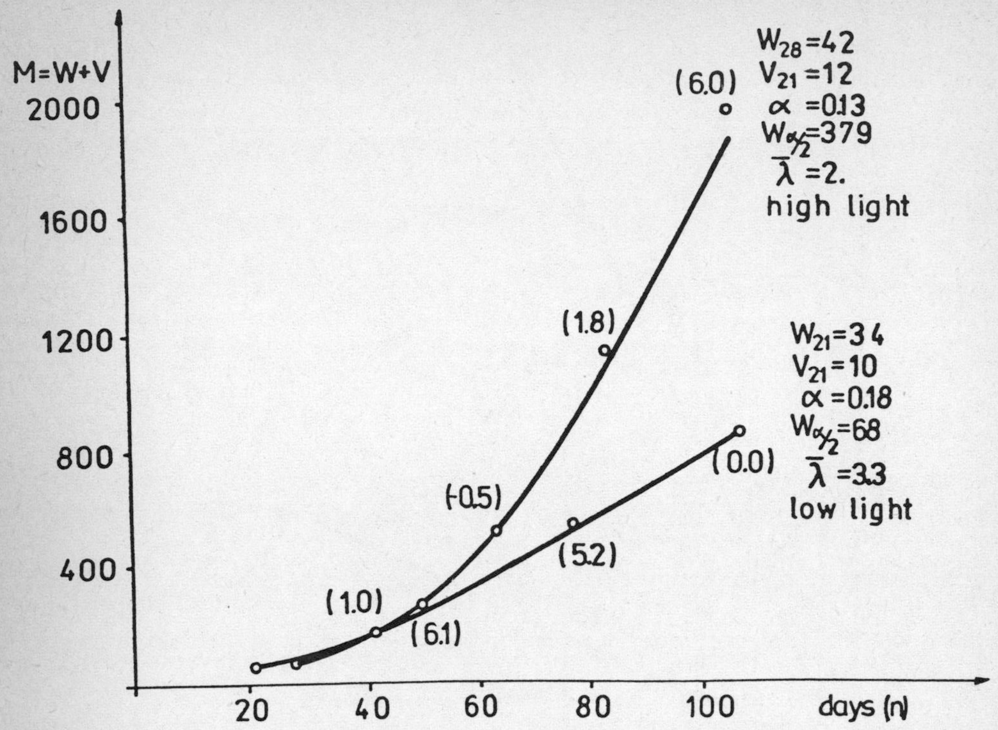

Fig. 3. Comparison between predictions of the model and mean experimental values from the same data sets as in Fig. 1 and 2. W_n and V_n - initial weights of assimilatory and nonassimilatory parts, respectively; α-unit leaf rates at the beginning of growth measurements (estimated roughly from experimental data and fitted exactly by computer iterations); $W_{\alpha/2}$-weight of assimilatory organs at which α decreases to 1/2 (fitted by computer iterations); $\bar{\lambda}$-ratio of assimilatory to nonassimilatory parts as estimated from experimental data at the end of the growing season; in brackets the (specific) relative error of particular measurements.

REFERENCES

[1] Charles-Edwards, D.A.: Physiological determinants of Crop Growth, Academic Press, London, 1982.

[2] Evans, G.C.: The quantitative analysis of plant growth. Blackwell Scientific Publications, Oxford, London, Edinburgh, Melbourne, 1972.

[3] Thornley, I.H.M.: Mathematical Models in Plant Physiology, Academic Press, London, 1976.

[4] deWitt, C.T., et.al.: Simulation of assimilation, respiration and transportation of crops, Simulation Monographs, Pudoc, Wageningen, 1978.

[5] Żelawski, W., and Szlenk, W.: A concept of vegetative growth and a dynamic model of dry matter accumulation in plants on an example of Scots pine seedlings, III Internationales Symposium Biophysik der Pflanzen, Reinholdsbrunn 1983, Zeitschrift der Humboldt-Universität, Berlin 1984 (in print).

[6] Żelawski, W., and Szlenk, W.: A concept and a dynamic model of vegetative growth in plants, Physiologia Plantarum (in print).

[7] Szlenk, W., and Żelawski, W.: "Plant growth as a dynamical system", preprint No. 7 1984/85 Matematisk Institut, Aarhus Universitet.

W. Szlenk
Institute of Mechanics
University of Warsaw
PKiN, IX p.
PL-00-901 Warsaw, Poland

W. Żelawski
Institute of Plant Biology
Warsaw Agricultural University
ul. Rakowiecka 30/34
PL-00-901 Warsaw, Poland

PHANTOM ITERATES OF CONTINUOUS FUNCTIONS

György Targonski

1. THE PROBLEM OF ITERATIVE ROOTS

For a self-mapping f of a set S, the functional equation

$$g^r = f \quad (r \in \mathbb{N},\ r \geq 2) \tag{1.1}$$

defines an r-th iterative root. In general, no such mapping exists and the theory becomes extremly complicated when f is not bijective; the general problem of the non-bijective case remained unsolved until 1977 (Isaacs 1950, Łojasiewicz 1951, Sklar 1969, Zimmermann 1976, 1978, see also Targonski 1981, section 2.1). The mathematician's traditional reaction to such a situation is trying to define "generalized", "improper" iterative roots, for example by extending f to a set larger than S in an appropriate way. (See: Bajraktarevič 1965, Mira, Müllenbach 1983). Iterative roots in this latter sense are still mappings, although of a different set.

We propose a different approach which we present on three different levels of abstraction, giving an explicit construction on the third, most concrete level.

2. PHANTOM ITERATES

The most abstract approach to generalized iterative roots is the one usual in such a situation, the best-known example being the way imaginary numbers are introduced as square roots of negative reals. The semigroup under composition of self-mappings of a set is isomorphically immersed in a larger semigroup where a solution for the equation corresponding to (1.1) may exist.

The middle level is the one on which we give the definition of phantom iterate. Let A be a commutative algebra, and consider a family Φ of functions from S into A, closed under the operations of A. Thus the elements of Φ are functions with independent variable in S and dependent variable in A. Under the operations of A: "pointwise" addition, multiplication, and multiplication by a scalar, Φ becomes itself a commutative algebra. Then the substitution operator (right composition operator) T as defined by

$$T\phi := \phi \circ f \quad \phi \in \Phi \tag{2.1}$$

is an endomorphism of Φ. In particular, T is a linear transformation on Φ considered as a vector space, that is,"forgetting" multiplication. Assume now that f has no r-th iterative root for a given r, but there exists a linear transformation U on Φ as a vector space (not necessarily an endormophism of Φ as an algebra) such that $U^r = T$. Then, U is a "phantom r-th iterative root" of f:

(2.2) Definition. For given f and r, let the linear transformation U be such that $U^r\phi = \phi \circ f$ for all $\phi \in \Phi$. We call U a phantom iterative root of f (of order r).

In analogy with Definition 2.2 we can also define phantom continuous iterates (phantom embeddings):

(2.3) Definition. For given f let the one-parameter semigroup of linear transformation $\{U^t, t > 0\}$ be such that $U^1\phi = \phi \circ f$ for all $\phi \in \Phi$, (therefore $U^n\phi = \phi \circ f^n$, $n \in \mathbb{N}$). We call $\{U^t, t > 0\}$ a phantom continuous iterate (phantom $\mathbb{R}^+$-embedding) of f

3. CONTINUOUS SELF-MAPPINGS OF A COMPACT INTERVAL

We carry out our construction for C(I,I), the set of all continous self-mappings of a compact real interval I. Any substitution operator on the space C(I)

$$(T\phi)(x) := \phi[f(x)] \quad f \in C(I,I),\ \phi \in C(I) \tag{3.1}$$

is, as stated above in a more general setting, an endomorphism, but in this particular case the opposite is also true.

(3.2) Proposition. Every endomorphism T of the algebra (under point-wise multiplication) C(I) is of the form $T\phi = \phi \circ f$ ($f \in C(I,I)$) that is,

$$\forall\ \phi_1,\phi_2\ T\lambda\phi_1 = \lambda T\phi_1\ \&\ T(\phi_1+\phi_2) = T\phi_1+T\phi_2\ \&\ T(\phi_1\phi_2) = T\phi_1 T\phi_2 \Leftrightarrow$$

$$\Leftrightarrow\ \exists f\ \forall\phi\quad T\phi = \phi \circ f \qquad \phi_1,\phi_2 \in C(I),\ f \in C(I,I),\ \phi \in C(I).$$

This follows from Proposition 5.1.7 in Targonski 1970. (Cf.also Targonski 1981, pp.137-138).

This fact renders the task of isomorphic immersion in a larger semigroup more elegant: the substitution operators are precisely the endomorphisms of the algebra Φ, and their semigroup is immersed in the vector space of all linear transformations on Φ (considered as a semigroup under multiplication).

In order to show the existence of phantom roots one has to show the existence of $U = T^{1/r}$ as given by Definition 2.2. (Obviously, T is bounded and $||T|| \leq 1$.) We will not concern ourselves here with the general problem. Rather, in the next section, we shall exhibit a class of phantom square roots.

4. EXPLICIT CONSTRUCTION OF A CLASS OF PHANTOM SQUARE ROOTS

In this section $I = [0,1]$, moreover we restrict our attention to the set $C_0(I,I)$ and to the space $C_0(I)$, comprising precisely those elements of $C(I,I)$, respectively of $C(I)$ having a fixed point at zero. (We could drop this condition and formulate the results for $C(I)$; see Problem IV below.)

We find that on $C_0(I)$, the zero operator is a substitution operator. Denote by $0 \in C_0(I,I)$ the mapping $0(x) \equiv 0$; we find that $(\phi \circ 0)(x) = \phi[0(x)] \equiv \phi(0) = 0 \equiv 0(x)$, right composition with 0 actually gives rise to the zero operator on $C_0(I)$.

We now exhibit a continuous self-mapping of the closed unit interval with a phantom square root.

Let $\alpha,\beta \in C_0(I,I)$ with the following properties

$$\alpha(x)\beta(x) \not\equiv 0 \tag{4.1}$$

$$\begin{aligned}
&\text{(a)} \quad \alpha[\alpha(x)]\ \beta[\alpha(x)] \equiv 0 \\
&\text{(b)} \quad \alpha[\alpha(x)]\ \alpha[\beta(x)] \equiv 0 \\
&\text{(c)} \quad \alpha[\alpha(x)]\ \beta[\beta(x)] \equiv 0 \\
&\text{(d)} \quad \beta[\alpha(x)]\ \alpha[\beta(x)] \equiv 0 \\
&\text{(e)} \quad \beta[\alpha(x)]\ \beta[\beta(x)] \equiv 0 \\
&\text{(f)} \quad \alpha[\beta(x)]\ \beta[\beta(x)] \equiv 0
\end{aligned} \tag{4.2}$$

We introduce the notations

$$A\phi := \phi \circ \alpha \qquad B\phi := \phi \circ \beta \qquad T := (A + B)^2 \tag{4.3}$$

(4.4) Proposition. The mapping f defined by $f(x) = \alpha[\alpha(x)] + \beta[\alpha(x)] + \alpha[\beta(x)] + \beta[\beta(x)]$ has a phantom iterative root given by $T^{1/2} := A + B$ and thus determined by the pair of mappings (α,β). The phantom root $A + B$ is not a substitution operator.

Proof. $A + B$ is not a substitution operator, since it is not an endomorphism. Note that for the identity mapping x on [0,1], $x \in C_0(I)$. If $A + B$ were an endomorphism, then in particular $(A + B)x^2 = \alpha(x)^2 + \beta(x)^2 = [(A + B)x]^2 = [\alpha(x) + \beta(x)]^2 = \alpha(x)^2 + 2\alpha(x)\beta(x) + \beta(x)^2$, therefore $\alpha(x)\beta(x) \equiv 0$. From (4.1) now follows that $A + B$ is not a substitution operator.

On the other hand, $(A+B)^2$ is a substitution operator. First of all,

$$(A+B)^2x = (A^2+AB+BA+B^2)x=\alpha[\alpha(x)]+\beta[\alpha(x)]+\alpha[\beta(x)]+\beta[\beta(x)]=:f(x) . \qquad (4.5)$$

We note that because of (4.2), for any x at most one of the four terms is different from zero, that is, there exists a disjoint decomposition $I = I_{11} \cup I_{21} \cup I_{12} \cup I_{22}$ such that

$$f(x)=\begin{cases}\alpha[\alpha(x)] & x \in I_{11}\\ \beta[\alpha(x)] & x \in I_{21}\\ \alpha[\beta(x)] & x \in I_{12}\\ \beta[\beta(x)] & x \in I_{22}\end{cases} \qquad (4.6)$$

We now show that

$$(A + B)^2x^n = x^n \circ f \qquad (4.7)$$

for all $n \in \mathbb{N}$; by definition (see (4.5))this is true for n = 1. From (4.7),

$$(A+B)^2x^n=(A^2+BA+AB+B^2)x^n=\alpha^2(x)^n+\beta[\alpha(x)]^n+\alpha[\beta(x)]^n+\beta^2(x)^n \quad . \qquad (4.8)$$

Because of (4.2), we also have

$$(A+B)^2x^n=[(A+B)^2x]^n=(\alpha[\alpha(x)]+ \beta[\alpha(x)]+\alpha[\beta(x)]+\beta[\beta(x)])^n , \qquad (4.9)$$

as comparison with (4.8) shows, the "mixed" terms vanish because of (4.2).

From (4.9) and the linearity of $(A + B)^2$ it follows that for every polynomial $p \in C_0(I)$,

$$(A+B)^2p = Tp = p \circ f = T_fp \text{ where } T_f\phi : = \phi \circ f \ (\phi \in C_0(I)) . \qquad (4.10)$$

Let now $\phi \in C_0(I)$ be given, $\|\phi\| = 1$ and $\varepsilon > 0$ then the Weierstrass approximation theorem assures the existence of a polynomial $p \in C_0(I)$, $|\phi(x) - p(x)| < \frac{\varepsilon}{5}$. Now $\|(A+B)^2\| \le \|A^2\| + 2\|A\|\cdot\|B\| + \|B^2\| \le 4$, $\phi(x) = p(x) + \eta(x)$ ($|\eta| \le \frac{\varepsilon}{5}$) and $((A+B)^2-T_f)\phi=$ $= [(A+B)^2 - T_f][p(x) + \eta(x)]$. Because of (4.10), this reduces to $((A+B)^2-T_f)\phi =$ $= ((A+B)^2-T_f)\eta$. Then $\|((A+B)^2-T_f)\phi\| \le \|((A+B)^2-T_f\|\cdot\frac{\varepsilon}{5} \le \frac{5\varepsilon}{5} = \varepsilon$ for all $\varepsilon > 0$ and all ϕ with $\|\phi\| = 1$.

Thus in $C_0(I)$ (cf.(4.3))

$$(A + B)^2 = T = T_f \tag{4.11}$$

thus T is a substitution operator. It is not excluded for T also to have a square root which is an endomorphism, that is, one which is a substitution operator. Then, beside the phantom, f also has a "true" square root. We shall later see such a case.

We now describe a class which serves as example. Let $\alpha,\beta \in C(I,I), I_1 := [0,\frac{1}{3}]$, $I_2 := [\frac{1}{3},\frac{2}{3}]$, $I_3 := [\frac{2}{3},1]$

$$\alpha(x)\begin{cases} \equiv 0 & x \in I_1 \\ \not\equiv 0, \leq \frac{2}{3} & x \in I_2 \\ \not\equiv 0, \leq \frac{1}{3} & x \in I_3 \end{cases}$$
$$\beta(x)\begin{cases} \equiv 0 & x \quad I_1 \cup I_2 \\ \not\equiv 0, \leq \frac{2}{3} & x \in I_3 \end{cases} \tag{4.12}$$

and (cf. (4.1))

$$\alpha(x)\beta(x) \not\equiv 0 \qquad (x \in I_3) \quad . \tag{4.13}$$

The behaviour in the three intervals of the four terms in (4.5) is given by the following table

(4.14)

	I_1	I_2	I_3
$\alpha[\alpha(x)]$	$\equiv 0$	$\not\equiv 0$	$\equiv 0$
$\beta[\alpha(x)]$	$\equiv 0$	$\equiv 0$	$\equiv 0$
$\alpha[\beta(x)]$	$\equiv 0$	$\equiv 0$	$\not\equiv 0$
$\beta[\beta(x)]$	$\equiv 0$	$\equiv 0$	$\equiv 0$

It follows from (4.14) that (4.2) is satisfied, and so is (4.1) by (4.13). So we have an uncountable class of functions with phantom square roots. In this case, (4.5) reduces to

$$f(x) = \alpha[\alpha(x)] + \alpha[\beta(x)] \quad . \tag{4.15}$$

Let us conclude this section by a rather simple-minded example of a mapping with both a true and a phantom square root. As seen above, the zero operator is a substitution operator; it has itself as square root, that is, the zero mapping has itself as "true" square root. Consider on the other hand

$$\alpha(x),\beta(x) \begin{cases} \equiv 0, & 0 \le x \le \frac{2}{3} \\ \not\equiv 0, & \le \frac{2}{3}, \ \frac{2}{3} \le x \le 1 \end{cases} \quad . \tag{4.16}$$

It is possible to choose α,β in such a way that (cf. (4.1))

$$\alpha(x)\beta(x) \not\equiv 0 \qquad \frac{2}{3} \le x \le 1 \quad . \tag{4.17}$$

From a table constructed in analogy with (4.14), one finds that now

$$\alpha[\alpha(x)] \equiv \beta[\alpha(x)] \equiv \alpha[\beta(x)] \equiv \beta[\beta(x)] \equiv f(x) \equiv 0 \ , \tag{4.18}$$

thus $A + B \not\equiv 0$ is a phantom square root of the zero operator. Because of (4.17), it is not a "true" square root.

5. PROBLEMS

I. The first question is of course under what conditions α,β exist (in a unique way?) for given f, so that a representation (4.6) holds with all the consequences sketched.

II. One could investigate phantom square roots given as sum of three or more substitution operators.

III. For phantom roots of order r, a similar approach could be attempted.

IV. The results can be extended to C(I) if we drop the condition that the zero operator be a substitution operator. In fact, this condition can be considered somewhat artificial, in view of the Theorem on p. 204 in Katznelson 1968, which plays a crucial part in the proof of Proposition 5.1.7 in Targonski 1970.

V. A more ambitious program (it motivated this investigation) is to construct a phantom continuous iterate. One would then have a continuous time phantom (semi-) dynamical system.

VI. One can ask (this was suggested by L. Reich) whether phantom roots could be used to construct "true" roots, provided they exist. This is possible in certain cases, as can be seen if a result in Reich, Schwaiger 1980 is expressed in our present terminology. (See also W. Hösler's result in Targonski 1984, p. 25).

REFERENCES

Bajraktarević, M.: Solution général de l'équation fonctionnelle.. Publ. Inst. Math. Beograd (N.S.) 5 (19)(1965), 115-124.

Isaacs, R.: Iterates of fractional order. Canad. J. Math. 2 (1950), 409-416.

Katznelson, Y.: An Introduction to Harmonic Analaysis, Wiley 1968.

Łojasiewicz, S.: Solution général de l'équation fonctionelle... Ann.Soc.Polon. Math. 24 (1951), 88-91.

Mira, C., Müllenbach, S.: Sur l'iteration fractionnaire d'un endomorphisme quadratique. C.R. Acad. Sci. Paris 297 (1983), 369-372.

Reich, L., Schwaiger, J.: Eine Linearisierungsmethode für Funktionalgleichungen von iterativem Typus in Potenzreihenringen. Aequ.Math.20 (1980), 224-243.

Sklar, A.: Canonical decompositions, stable functions and fractional iterates. Aequ. Math. 3 (1969), 118-129.

Targonski, Gy.: Linear endomorphisms of function algebras and related functional equations. Indiana Univ. Math. J. 20 (1970), 579-589.

Targonski, Gy.: Topics in Iteration Theory. Vandenhoeck & Ruprecht, Göttingen/ Zürich, 1981.

Targonski, Gy.: New directions and open problems in iteration theory. Ber. math.-stat. Sekt. Forschungszentrum Graz, Nr. 229 (1984).

Zimmermann (née Riggert), G.: n-te iterative Wurzeln von beliebigen Abbildungen. Lecture at the 1975 International Symposium on Functional Equations. Abstract in Aequ. Math. 14 (1976), 208.

Zimmermann, G.: Über die Existenz iterativer Wurzeln von Abbildungen. Doctoral Dissertation, University of Marburg, 1978.

György Targonski
Fachbereich Mathematik
Universität Marburg
Lahnberge
D-3550 Marburg
Federal Republic of Germany

COMPETITION BETWEEN ATTRACTIVE CYCLE AND STRANGE ATTRACTOR

Roger Thibault

1. For the past few years it has been recognized that the most interesting features of recurrence equations (i.e. iterative systems) such as the appearance of strange attractors (Hénon's attractor, 1976), may be obtained by using peacewise linear functions (Lozi, 1978). Misiurewicz has given a mathematical proof of the chaotic behaviour in such a case (1980).

2. Generally, the transformation studied has a constant Jacobian J (J = 1 in the conservative case). The model used here introduces a coefficient playing the role of damping. So the Jacobians J_1, J_2 of the two determinations of the transformation T are different, and the powers of T are associated with a Jacobian $J_1^{n_1}J_2^{n_2}$ which take infinitely many values when n_1, n_2 are varying.

Let T $x_{n+1} = \alpha x_n + y_n \quad , \quad y_{n+1} = -x_n + \lambda|y_n| - 1$ be a second order

recurrence, where the value of λ is fixed, the main parameter being α. The two determinations of T as linear transformations are denoted $T_1(y > 0)$ and $T_2(y < 0)$. λ is chosen to reduce the conservative case ($\alpha = 0$) to the simplest one: namely the value $\sqrt{2}$ insures the reduction of T to a root of the identity when the variables tend to infinity ($T^{16} \to I$).

3. Let us begin by the study of the conservative case $\alpha = 0$. The fixed point $x = y = \sqrt{2}/2 - 1$ is a center, and in the vicinity of it, T^8 reduces to identity.

The research of the cycles of a given order is facilitated by the following property of symmetry with respect to the first bisectrix: if two points A_n, B_n are symmetrical, the consequent A_{n+1} of A_n and the antecedent B_{n-1} of B_n are also symmetrical. The cycles are of two species:

- either including a point belonging to the first bisectrix
- or including two consecutive points symmetrical with respect to the first bisectrix.

This remark enables one to manage a systematic research of all cycles in the conservative case. Analytic treatment is possible for low values of the order, numerical treatment for others. Two cases are studied:

1. Search for an initial point $x_0 = y_0$ such that

a) $y_n = y_{n+1}$ or

b) $y_n = y_{n+2}$

2. Search for an initial point whose the antecedent is symmetrical, i.e. (for the positive case) $2y_0 = \lambda x_0 - 1$ with conditions a) or b) above.

Results. After noting the existence of a continuous family of cycles of order 8 (containing as a particular case a cycle of order 4), let us particularly consider the double sequence of cycles of order 5 k+1 (k = 3,4...). k = 3 corresponds to cycles at infinity already noted (homogeneous case). For k > 3, if k is odd, the two cycles are obtained following the procedure 1 b), one for $x_0 > 0$, the other for $x_0 < 0$. If k is even, one cycle is given by 1a), the other by 2). The cycles of order 5, of type $T_1^4 T_2$ and $T_1^3 T_2^2$, may be interpreted as limiting case when $k \to \infty$.

The genus of this cycle has to be specified for the following study: of the two cycles of order 5, the first is a center, the second a saddle. The same result can be observed for the pairs of (5k+1)-cycles for k = 4,5,6:

order	generated by	genus
21	$x_0 = y_0 = 66{,}23$	saddle
21	$x_0 = y_0 = -9{,}95$	center
26	$x_0 = y_0 = 46{,}33$	center
26	$x_0 = 1.59$ $y_0 = 2.96$	saddle
31	$x_0 = y_0 = 38{,}09$	center
31	$x_0 = y_0 = -3.28$	saddle

The two following values of k give only cycles of genus saddle (cycles of order 36 and 41).

4. Dissipative case. Small damping, α is assumed to be negative and sufficiently near of 0. Cycles may continuously follow starting from the conservative case. Their genus is preserved if it is a saddle, and becomes an attractive focus in the case of a center. When α decreases, the cycles disappear progressively, by coalescence of two points on the transition line y = 0, one of them belonging to a focus cycle, the other to a saddle-cycle of the same order (only the order 5, 21, 26, 31 have been examinated here). Corresponding values of α are

k = 3	(cycle of order 16 at infinity)		α = 0
k = 4	cycle of order 21		α = -0.0231
k = 5		26	α = -0.0154
k = 6		31	α = -0.0034
	cycle of order	5	α = -0.4476

For an attractive focus cycle, the attraction domain in $\mathbb{D}$ is now considered. The value $\alpha = -0.022$ is retained for numerical computation, then two cycles do exist as attractive focuses, of order 5 and 21.

Consider the point P_o ($X_o < 0$, $Y_o > 0$) belonging to the attractive 5-cycle. The attraction domain contains firstly a "proximal domain" $\mathcal{D}_o$ whose points are of the same "type" (i.e. involve the same sequence of transformations T_i (i = 1 or 2) as the cycle itself) and converge directly using the linear transformation $T^5 = T_1{}^4 T_2$. $\mathcal{D}_o$ is bounded by the broken line obtained by the inverse iterate of T^5 of the transition line, as long as T^5 remains of the type $T_1{}^4T_2$. The attraction domain is thus obtained by the union $\mathcal{D} = \bigcup_{p=1,+\infty} T^p(\mathcal{D}_o)$ where T is the transformation T^{-5}.

Let $P_1(X_1 < 0, Y_1 > 0)$ the point belonging to the 5-cycle of genus saddle. One invariant line (repulsive) issued of P_1 enter into $\mathcal{D}_o$; so this invariant line belongs to $\mathcal{D}$. The invariant line (attractive) L_1' directed toward y > 0 plays the role of boundary for $\mathcal{D}$. Successive inverse iterations give a broken line with the shape of diverging spiral. Unfortunately, the invariant line (attractive) issuing of P_1 in the opposite direction L_1' has some homoclinic points, i.e. it intersects the invariant line (repulsive) issuing of the point P_5, antecedent of P_1 by T. So there are infinitely many points in the vicinity of P_5 and L_1' is very complicated (see Fig.2). This gives $\mathcal{D}$ a fuzzy boundary.

Nevertheless, it has been observed that after a transitory behaviour, all initial conditions belonging to the domain limited by the first segments of L_1, L_1' and their ^{4}iterates are converging toward the attractive 5-cycle. This fact will be explained later.

Similar results are obtained for the 21-cycle, the domain $\mathcal{D}_o'$ being very small (for $\alpha = -0.022$) (α is close to the value of disappearing of the cycle). The domain in $\mathcal{D}'$ is of spiral form, imbricated with the domain $\mathcal{D}$.

5. Dissipative case. Strange attractor. For the value $\alpha = -.3$, numerical computation seems to show the appearance of a strange attractor. It has been proved in [4] that there exists an absorbing domain (i.e. a domain D such that $T(D) \subset D$) and no attractive cycle previously studied inside. The boundary between the attraction bassin of this strange attractor and the domain of the attractive 5-cycle is shown on Fig.3. The attraction bassin has five corridors, of spiral shape, and going to infinity.

The question of determining the value of α for which the attractor is born has been solved in [4]. It is connected to the relative position of two invariant lines, the first L_1 (repulsive) issuing of the point P_1 belonging to the 5-cycle of saddle-genus, the second L_2 (attractive) issuing of the antecedent of P_1. Existence of intersection points of this two lines prevents the construction of the absorbing domain D: iterative process in the chaotic region allows a point to escape when it falls into the triangle sketched on Fig.4. It is then attracted by the 5-cycle. This is the explanation of the fact mentioned at the end of § 4.

So the threshold value of α is obtained when the first angular point of L_1 falls on L_2. This value is a root of an algebraic equation, but it is most easily given by direct numerical computation: $\alpha_0 = -0.23548$.

One may see the deformation of the attractor when α is decreasing from the above value α_0. It remains alone after the vanishing of the 5-cycle for $\alpha = -0.4476$. It has been identified down to the value $\alpha = -1.5$, then seems to become unstable roughly for $\alpha = -1.6$. This problem requires further investigation.

FIGURES :

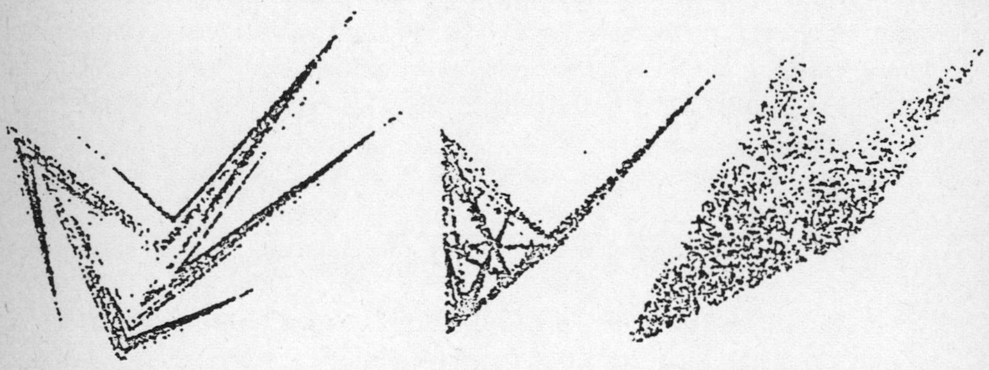

$\alpha = -.3$ $\alpha = -1$ $\alpha = -1.4$

Evolution of the attractor

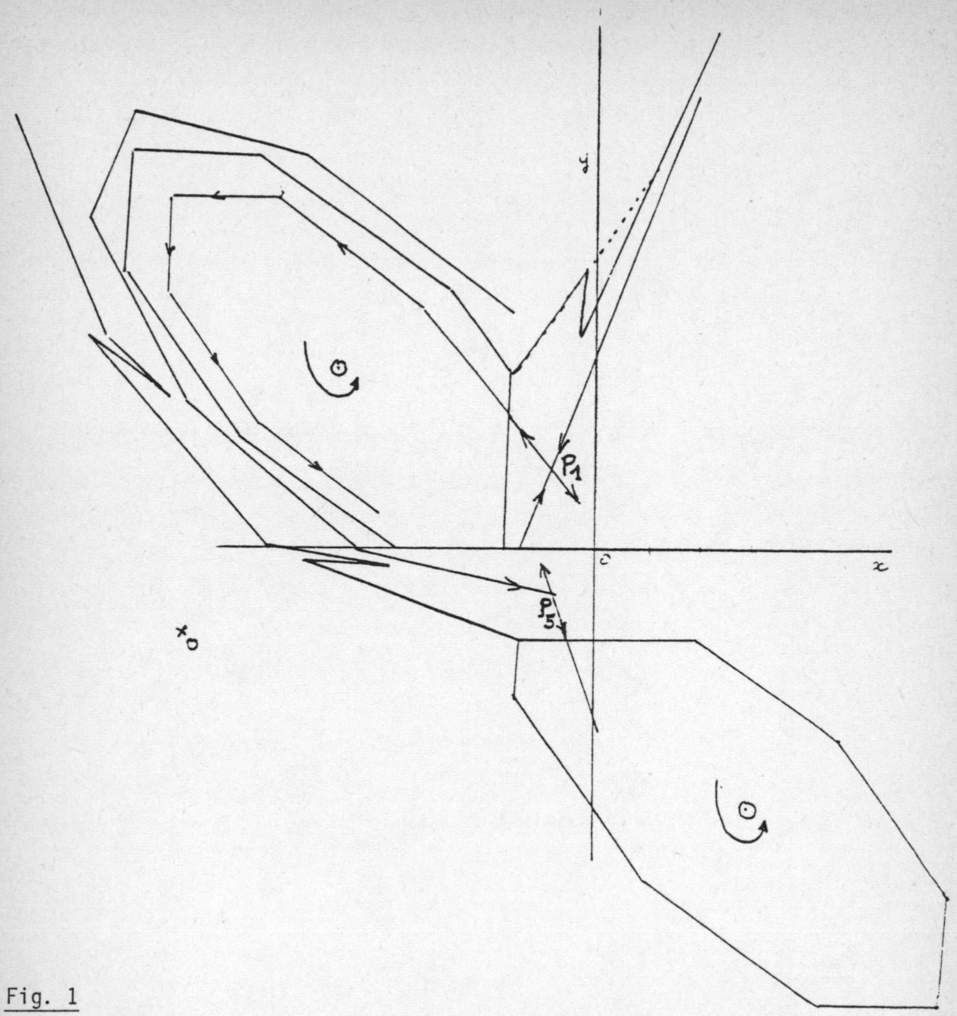

Fig. 1

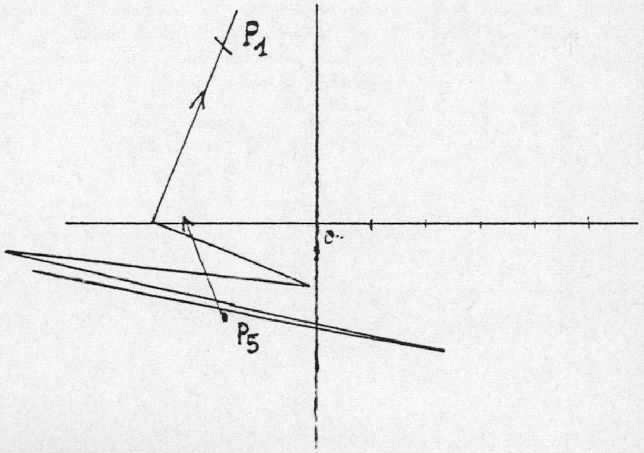

Fig. 2

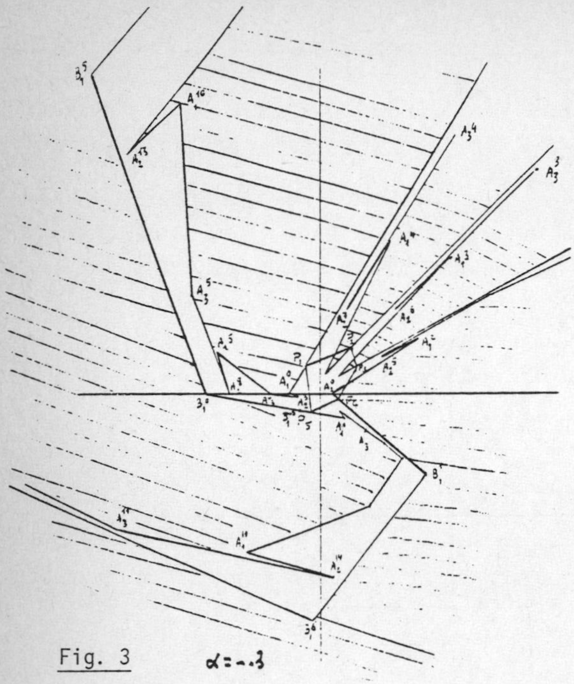

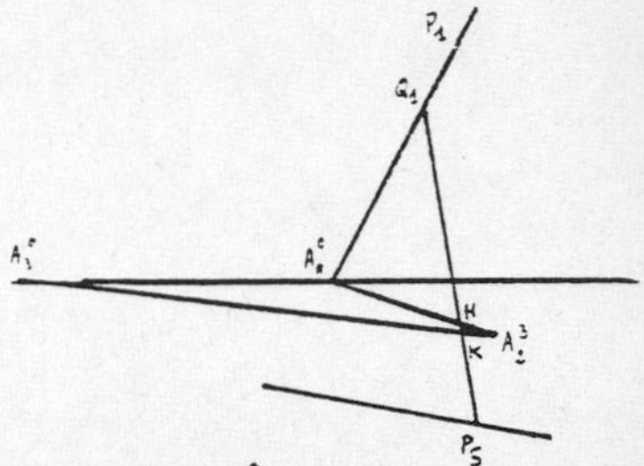

$\alpha = -.2$

Fig. 3 $\alpha = -.3$

REFERENCES

[1] Hénon, M.: A two-dimensional mapping with a strange attractor. Commun.Math. Phys.50, 69-77 (1976).

[2] Lozi, R.: Un attracteur étrange (?) du type attracteur de Hénon. J.Phys.(Paris) 39 (Coll.C5), 9-10 (1978).

[3] Misiurewicz, M.: Strange attractors for the Lozi mappings in R.M.G.Helleman (ed.) Nonlinear Dynamics (1980).
Annals of the New York Academy of Sciences 357, 348-358 (1980).

[4] Thibault, T.: Naissance d'un attracteur (1984).
Note C.R.Acad.Sci. Paris, A299, p.943 (1984).

Roger Thibault
Laboratoire de Mécanique
Université Paul Sabatier
31062 Toulouse
France

ON THE RELATION BETWEEN ORBITS OF AN ITERATION SEMIGROUP AND THE ORBITS OF THE EMBEDDED MAPPINGS

Jürgen Weitkämper

It is known (cf. Zimmermann 1978, Targonski 1981) that orbits of iterative roots are unions of orbits of the given mapping. Especially cycles of the root are unions of cycles. Zimmermann 1978 describes relations between the periods of the cycles of the given mapping and the root cycle. As iterative roots are a first step in the problem of embedding a mapping in a "continuous" iteration semigroup we try to generalize these results to "continuous" orbits hoping to find in this way new criteria for embeddability. The "continuous" orbits turn out, under not too severe conditions, to consist of exactly one orbit or to be unions of uncountably many orbits of the embedded mapping. In the former case the orbit contains a stationary point. We discuss the form of splinters on closed trajectories and give formulas that put the period of the closed trajectory in relation to the period of the embedded cycles. All results are stated for semigroups but are valid for groups as well.

1. DEFINITIONS

Let f be a self-mapping of a set X. The natural iterates f^n of f are recursively defined by $f^0 = Id$, $f^{n+1} = f \circ f^n$. The splinter $S_f(x)$ is the set $\{f^n(x);\ n \in \mathbb{N}_0\}$ and a k-cycle $\{x_1,\ldots,x_k\}$ consists of k different points with $f(x_i) = x_{i+1}$, $i = 1,\ldots,k-1$ and $f(x_k) = x_1$. Each point of a k-cycle is called k-periodic for f. An orbit $\Omega_f(x)$ is the equivalence class of x under the relation $x \sim_f y :\Leftrightarrow \exists\ n,m:$ $f^m(x) = f^n(y)$. The natural iterates form a semigroup, since $f^n \circ f^m = f^{n+m}$ for all n and m. This leads to the following generalization. A family $\{f^t;\ t \geq 0\}$ of self-mappings of a set is called an iteration semigroup if $f^s \circ f^t = f^{s+t}$ for all $s,t \geq 0$. An iteration semigroup $\{f^t\}$ is called an embedding of f if $f^1 = f$. In analogy to the notions of splinters and orbits we define the trajectory of x to be the set $T(x) = \{f^t(x);\ t \geq 0\}$ and the (continuous) orbit $O(x)$ to be the equivalence class of x under the relation $x \sim y :\Leftrightarrow \exists\ s,t : f^s(x) = f^t(y)$. If $f^t(x)=x$ for some $t>0$ the trajectory $T(x)$ is called closed and a stationary point z satisfies $f^t(z) = z$ for all $t \geq 0$. Clearly $\Omega_{f^t}(x) \subset O(x)$ and $S_{f^t}(x) \subset T(x)$ for all x and t. An f-orbit contains at most one cycle and a continuous orbit at most one closed trajectory.

2. THE SEMIGROUP OF PERIODS

An orbit contains a closed trajectory if the mapping $t \to f^t(x)$ is not injective. For let $s,t(s < t)$ be two real numbers with $f^s(x) = f^t(x) =: z$. Then $T := t-s > 0$ and $f^T(z) = f^{t-s}(f^s(x)) = f^t(x) = z$. For a point x we define the set $P(x) :=$ $\{t > 0;\ f^t(x) = x\}$. Thus P(x) being not empty is equivalent to T(x) being a closed trajectory.

2.1. Lemma

Let P(z) be not empty. Then
(a) P(z) is a semigroup (under addition),
(b) if $s,t \in P(z)$, $s < t$, then $t-s \in P(z)$,
(c) $P(y) = P(z)$ for all y in the trajectory of z.

We call P(z) the semigroup of periods and because of (c) it is justified to speak of the semigroup of periods of a closed trajectory.

Proof of 2.1 (a) Let $s,t \in P(z)$. Then $f^{s+t}(z) = f^s(f^t(z)) = f^s(z) = z$, thus $s+t \in P(z)$. (b) Now let $s,t \in P(z)$, $s < t$. Then $f^{t-s}(z)$ is defined, and $f^{t-s}(z) = f^{t-s}(f^s(z)) = f^t(z) = z$. Thus $t-s \in P(z)$. (c) Let $y = f^s(z)$ be an element of the trajectory T(z), and $t \in P(z)$. Then $f^t(y) = f^{t+s}(z) = f^s(f^t(z)) = f^s(z) = y$, thus $P(z) \subset P(y)$. The other inclusion is proved using (a).

One expects that for a closed trajectory not consisting of a stationary point, there exists a smallest time $t_0 > 0$ in P(z) being the "time of first return" of the state z. That this is not always true shows the following example of a "microperiodic" time-dependence $t \to f^t(x)$.

2.2. Example

Let B be a Hamel base for the real numbers containing 1, and $u:\mathbb{R} \to \mathbb{Q}$ be the group homomorphism that maps each real x onto the coefficient of 1 in its Hamel base expansion. The mappings $f^t(x) := e^{u(t)}x$ form an iteration semigroup on $\mathbb{R}$. For $x \neq 0$ the set P(x) is different from $\mathbb{R}^+$ and dense in $\mathbb{R}_0^+$.

2.3. Proposition (Sklar 1982)

If P(z) is not empty then exactly one of the following possibilities holds:
(a) $P(z) = \mathbb{R}^+$ (z is a stationary point),

(b) $P(z) \neq \mathbb{R}^+$, but $P(z)$ is dense in the non-negative reals,

(c) $P(z)$ contains a least element t_o. (t_o is then the generating element of the semigroup of periods.)

In the cases (a) and (c) we define the period of the closed trajectory to be 0 respectively t_o.

2.4. Definition

An iteration semigroup is called non-exceptional if each not empty $P(z)$ of a not stationary point z contains a least element.

In the case of continuous time dependence the exceptional case (b) is excluded:

2.5. Corollary

Let X be a Hausdorff space, and let $P(z)$ be not empty. If $(0,\infty) \ni t \to f^t(z)$ is continuous, then either z is a stationary point or $P(z)$ contains a least element.

For the rest of this paragraph we discuss closed trajectories with a period $t_o > 0$.

2.6. Lemma

Let $P(z)$ contain a least element $t_o > 0$. Then the mapping $[0,t_o) \ni t \to f^t(z) \in T(z)$ is bijective.

Proof. Let s,t be in $[0,t_o)$ with $f^s(z) = f^t(z)$ and assume $s < t$. Then $t-s < t_o$ and $y := f^t(z) = f^{t-s}(f^s(z)) = f^{t-s}(y)$, thus $t-s \in P(y) = P(z)$ (Lemma 2.1), in contradiction to the minimality of t_o.

Let $y = f^t(z)$ be in $T(z)$ and assume $t \geq t_o$. Let n be the largest integer with $nt_o \leq t$, then $t-nt_o < t_o$ and $f^{t-nt_o}(z) = f^{t-nt_o}(f^{nt_o}(z)) = f^t(z) = y$. Thus the mapping is surjective.

Combining (2.5) and (2.6) we have the expected:

2.7. Proposition

Let X be a Hausdorff space that satisfies the first axiom of countability. If $P(z)$ is not empty, $t \to f^t(z)$ is continuous, and $P(z)$ contains a least element t_o, then the trajectory $T(z)$ is homeomorphic to the circle S^1.

As a consequence, "one-dimensional" iteration semigroups on real intervals with the property that $t \to f^t(x)$ is continuous for all x in X (x in $f^0(x)$ is already sufficient) contain no closed trajectories with period different from 0, since there exist no subsets of $\mathbb{R}$ homeomorphic to S^1. Thus if a real mapping has a cycle of period ≥ 2, there exists no embedding in an iteration semigroup with the described properties. For embeddings in such semigroups cf. Zdun 1979, Targonski 1981, Weitkämper 1983.

3. THE RELATION BETWEEN THE PERIOD OF A CLOSED TRAJECTORY AND THE CYCLE PERIODS OF THE MAPPINGS f^t

3.1.Lemma

Let z be a periodic (aperiodic) point with respect to the mapping f^t, $t \neq 0$, and $s \neq 0$ with t/s rational. Then z is periodic (aperiodic) for f^s.

Proof. Let $t/s = p/q$ for integers p and q. i) Let z be aperiodic for f^t, i.e. $f^{nt}(z) \neq z$ for all natural numbers n. Assume there is some natural k with $f^{ks}(z) = z$. Then $z = f^{ks}(z) = f^{ktq/p}(z) = (f^{ktq/p})^p(z) = f^{kqt}(z)$, thus z is periodic for f^t-contradiction. ii) Let z be periodic for f^t, $z = f^{kt}(z)$ for some $k \in \mathbb{N}$. Then $z = f^{kt}(z) = f^{ksp/q}(z) = (f^{ksp/q})^q(z) = f^{kps}(z)$, thus z is periodic for f^s.

From now on we consider only non-exceptional iteration semigroups (Definition 2.4), i.e.especially semigroups with continuous time-dependence in view of Cor.2.5.

3.2.Proposition

Let P(z) contain a least element t_0, and let z be a k-periodic point of f^t, $t > 0$.

(a) Then z is periodic for f^s, $s > 0$, if and only if t/s is rational.

(b) If $t/s = p/q$, $p,q \in \mathbb{N}$, then the period k' of z for f^s is given by

$$k' = m/(k,\frac{s\cdot m}{t_0}), \quad \text{with} \quad m := kp/(kp,q) .$$

Proof. a) Due to Lemma 3.1 we only need to prove the "only if" part. Let z be k'-periodic for f^s : $z = f^{k's}(z)$. We have $k's \in P(z)$, $kt \in P(z)$, thus with Prop.2.3 there exist natural numbers n,n' with $nt_0 = kt$ and $n't_0 = k's$, thus

$$\frac{t}{s} = \frac{nt_0/k}{n't_0/k'} = \frac{nk'}{kn'} \in \mathbb{Q}.$$

b) Now let $t/s = p/q$, $p,q \in \mathbb{N}$. Then

$$f^{ms}(z) = f^{\frac{kp}{(kp,q)}s}(z) = f^{\frac{kqt}{(kp,q)}}(z) = z \ ,$$

since $\frac{q}{(kp,q)} \in \mathbb{N}$, and z is k-periodic for f^t.

Thus m is a multiple of the period k' of z for f^s, $ms \in P(z)$ and there exists an $n \in \mathbb{N}$ with $ms = nt_o$. Then $n = ms/t_o$ and $b = m/(m,sm/t_o)$ is defined. We have $f^{bs}(z) = f^{m/(m,n)\cdot nt_o/m}(z) = f^{nt_o/(m,n)}(z) = z$, and b is a multiple of the period of z for f^s. To finish the proof we have to show that b is the smallest number satisfying $f^{b\cdot s}(z) = z$. Assume there is an $i \in \mathbb{N}$ with $f^{si}(z) = z$, then $i\cdot s \in P(z)$ and $s\cdot i = wt_o$ for some $w \in \mathbb{N}$. Consequently $s/t_o = w/i = n/m = [n/(n,m)]/[m/(n,m)]$, thus $i \geq m/(n,m) = b$.

The last proposition shows how to compute the period k' for f^s from the period k for f^t, t_o and the ratio s/t. The period k' is not independent of t_o. For example a fixed point of the mapping f^1 can be an element of a closed trajectory with period $t_o = 1/n$ for each natural number n. For $s = 1/2$ (i.e. f^s an iterative square root of f^1) we have $k' = 2$ for $t_o = 1$ and $k' = 1$ for $t_o = 1/2$.

The next proposition gives a formula to compute the period k of a point z of an f^t-cycle from the ratio t/t_o (which has to be rational if z is periodic for f^t).

3.3. Proposition

Let P(z) contain a least element $t_o > 0$. Then

(a) z is periodic for f^s, $s > 0$ if and only if s/t_o is rational. If $t_o/s = p/q$, $p,q \in \mathbb{N}$, then the period is equal to $p/(p,q)$.

(b) z is aperiodic for f^s if and only if s/t_o is irrational.

Proof. (a) Since z is a fixed point for f^{t_o}, the first part is consequence of Prop.(3.2a). If we substitute t_o for t and 1 for k in (3.2b) we have $k' = m = p/(p,q)$ for the period of z for f^s.

With Lemma 2.6 a closed trajectory with a semigroup of periods that contains a least element $t_o > 0$ contains uncountably many points. If t/t_o is rational each point (cf.Lemma 2.1c) lies in a k-cycle of f^t. Thus the following theorem is proved:

3.4. Proposition

Let P(z) contain a least element t_o. If $t > 0$ with t/t_o rational, then f^t has uncountably many cycles of the period given in Prop.3.3.

The splinters of the mappings f^t for t/t_o irrational are dense in the closed trajectory:

3.5. Proposition

Let X be a Hausdorff space satisfying the first axiom of countability, P(z) contain a least element t_o, and let $t \to f^t(z)$ be continuous. If t/t_o is irrational, then the splinter $S_{f^t}(z)$ is dense in the closed trajectory T(z).

Proof. With (2.7) the closed trajectory T(z) is homeomorphic to the circle S^1. The assertion follows from the fact, that if $\beta \in \mathbb{R}$ and $\beta/2\pi$ is irrational, then $\{e^{in\beta}; n \in \mathbb{N}\}$ is dense in S^1.

4. CONSEQUENCES FOR THE PROBLEM OF EMBEDDING

Proposition (3.4) suggests that an orbit of an embedding is a union of uncountably many orbits of the embedded mapping. That is only partly true. The "continuous" orbit of a stationary point is identic with the "discrete" orbit.

4.1. Proposition

Let z be a stationary point of an iteration semigroup. Then

$$\Omega_{f^t}(z) = O(z) \text{ for all } t \neq 0 .$$

Proof. The inclusion $\Omega_{f^t}(z) \subset O(z)$ is trivial. Now let $y \in O(z)$, i.e. $f^u(y) = f^v(z)$ for some $u,v \geq 0$. Then $f^u(y) = z$, since z is a stationary point. Let n be an integer such that $nt > u$. Then $f^{nt}(y) = f^{nt-u}(f^u(y)) = f^{nt-u}(z) = z$, thus $y \sim_{f^t} z$ and $y \in \Omega_{f^t}(z)$.

From now on we consider only non-exceptional iteration semigroups that permit the definition of a period for each closed trajectory.

4.2. Theorem

Let O(z) be an orbit of an unexceptional iteration semigroup, $s \neq 0$. Then exactly one of the following cases occurs:

(a) O(z) contains a stationary point, then $O(z) = \Omega_{f^s}(z)$,

(b) $O(z)$ contains no closed trajectory, then $O(z)$ is the union of uncountably many acyclic f^s-orbits,

(c) $O(z)$ contains a closed trajectory with period $t_o > 0$, then if s/t_o is rational, then $O(z)$ is the union of uncountably many k-cyclic f^s-orbits with k given as in (3.3),

if s/t_o is irrational, then $O(z)$ is the union of uncountably many acyclic f^s-orbits.

Proof. Part (a) is Proposition (4.1).

The orbit $O(z)$ contains a closed trajectory if and only if the mapping $t \to f^t(z)$ is not injective.

(b) Let $t \to f^t(z)$ be injective. Then the splinters $S_{fs}(f^t(z))$ for $t \in [0,s)$ are pairwise disjoint, and thus are the f^s-orbits of the points $f^t(z)$. For let $y \in S_{fs}(f^u(z)) \cap S_{fs}(f^v(z))$, $u,v \in [0,s)$. Then there exist natural numbers n,m with $y = f^{ns}(f^u(z)) = f^{ms}(f^v(z))$, therefore $ns+u = ms+v$, since $t \to f^t(z)$ is injective. Further $|(n-m)s| = |v-u| < s$, which yields $u = v$. It is clear that $O(z)$ is the union of the orbits $\Omega_{fs}(f^t(z))$, $t \in [0,s)$.

(c) Let $O(z)$ contain a closed trajectory with period $t_o > 0$.

(1) For s/t_o rational $O(z)$ is the union of uncountably many cyclic f^s-orbits with Proposition (3.4) and Lemma (2.1c).

(2) Let s/t_o be irrational, z_o be an element of the closed trajectory, and k be the smallest natural number with $kt_o > s$. Then for all points y of the closed trajectory $T(z_o)$ we have $f^s \circ f^{kt_o-s}(y) = f^{kt_o}(y) = y = f^{kt_o-s} \circ f^s(y)$.

Thus the mapping $f^s|_{T(z_o)}: T(z_o) \to T(z_o)$ is bijective and the intersection $\Omega_{fs}(z) \cap T(z_o)$ of the "discrete" orbit $\Omega_{fs}(z)$ with the closed trajectory is a chain ordered like $\mathbb{Z}$ (it cannot be a cycle due to Proposition (3.3)). With Lemma (2.6) the closed trajectory contains a continuum of points, thus the orbit $O(z)$ has to be the union of uncountably many f^s-orbits.

The condition that we consider unexceptional iteration semigroups only is essential. In example (2.2) we defined an iteration semigroup on $\mathbb{R}$. Let x be not 0. Then $O(x) := \{y;\ \exists\, s,t \geq 0 : f^s(x) = f^t(y)\} = \{e^r x;\ r \in \mathbb{Q}\}$.

Proof. Suppose $y = e^r x$ for some $r \in \mathbb{Q}$. Then $f^0(y) = y = e^r x = e^{u(r)}x = f^r(x)$.

Thus $y \in O(x)$. Now let y be an element of $O(x)$, i.e. $f^s(y) = f^t(x)$ for some $s,t \geq 0$. Then $e^{u(s)}y = e^{u(t)}x$ and $y = e^{u(t)-u(s)}x = e^{u(t-s)}x$, thus $y \in \{e^r x; r \in \mathbb{Q}\}$.

In the same way one gets $\Omega_{f1}(x) = \{e^n x;\ n \in \mathbb{Z}\}$, thus $O(x)$ is a union of countably many f^1-orbits $\Omega_{f1}(f^r(x))$, where $0 \leq r < 1$ and r rational.

In connection with the problem of finding an iterative root of a mapping f the notion of "m-mateability" for a finite set of f-orbits was defined (Zimmermann 1978, Isaacs 1950. For a definition of "contraction", "curtailment", and "orbit isomorphism" we refer the reader to Chapter 1 and 2 in Targonski 1981.) On the union of m-mateable f-orbits it is possible to define an m-th iterative root with exactly one orbit. In the same way we now define "continuous mateability" for a set of f-orbits.

4.3.Definition

The f-orbits Ω_i, $i \in I$ are called continuously mateable if there exists an embedding of f in an unexceptional iteration semigroup on the union of these orbits that has exactly one (continuous) orbit.

From Theorem (4.2) we know that I consists of either just one or uncountably many indices. As for iterative roots (cf. Satz (1.10) and Satz (1.14) in Zimmermann 1978) we have the following two theorems:

4.4.Theorem

The mapping f is embeddable in an unexceptional iteration semigroup if and only if the set of all f-orbits permits a decomposition in subsets consisting of either exactly one or uncountably many orbits, such that the orbits in each class are continuously mateable.

4.5.Theorem

For the continuous mateability of f-orbits $\Omega_i, i \in I$ it is necessary that for each two orbits Ω_i, Ω_j there exists a contraction of Ω_i that is orbit isomorphic to a curtailment of Ω_j.

Proof. As a first step we show that there is a point z and a $\tau \in [0,1)$ such that $\Omega_i = \Omega(z)$ and $\Omega_j = \Omega(f^{\tau}(z))$. Let $x \in \Omega_i$, $y \in \Omega_j$. The two orbits Ω_i, Ω_j are contained in the same "continous" orbit thus there exist real numbers $s,u \geq 0$ with $f^s(x) = f^u(x)$. With a natural number $n > s$ we have $z := f^n(x) = f^{n-s}(f^s(x)) = f^{n-s+u}(y)$. Let m be the least natural number with $m \geq n-s+u$. Then $\tau := m-(n-s+u) \in [0,1)$ and $f^{\tau}(z) = f^{\tau}(f^{n-s+u}(y)) = f^m(y) \in \Omega(y)$. Therefore $\Omega_i = \Omega(z)$ and $\Omega_j = \Omega(f^{\tau}(z))$.

Define $H: \Omega_i \to \Omega_j$ by $H(x) = f^{\tau}(x)$. Clearly this map is an orbit homomorphism. If $H(x) = H(y)$, then $f(x) = f^{1-\tau}(f^{\tau}(x)) = f^{1-\tau}(f^{\tau}(y)) = f(y)$. Thus if we identify in Ω_i all points with the same image under H we get a contraction of Ω_i. The induced map $\bar{H}$ on this contraction is injective.

Now let $y = f(z) \in f(\Omega_j) \subset \Omega_j$, then with $x := f^{1-\tau}(z)$ we have $H(x) = f(z) = y$. Thus the only elements in Ω_j that have no preimage under H are first elements of Ω_j. Thus there exists a curtailment of Ω_j that makes $\bar{H}$ surjective.

The problem of embeddings in iteration semigroups would thus be solved if there could be found, as for m-mateable orbits, criteria for the continuous mateability of f-orbits.

Remark. For embeddings in iteration groups necessary and sufficient conditions are known (cf. Sklar 1978, Weitkämper 1983).

REFERENCES

Isaacs, R. 1950: Iterates of fractional order
Canad. J. Math. 2 (1950), 409-416.

Sklar, A. 1978: The embedding of functions in flows
Aequ. Math. 17 (1978), 367.

1982: Remark 6 presented by C. Alsina on the Symposium on functional equations in Oberwolfach 1982, Tagungsbericht 32/1982 des Math. Forschungsinstituts Oberwolfach.

Targonski, Gy. 1981: Topics in Iteration Theory
Vandenhoeck & Ruprecht, Göttingen, Zürich 1981.

Weitkämper, J. 1983: Embedding in Iteration Groups and Semigroups with Nontrivial Units
Stochastica, Vol.VII, No.3 (1983), 175-195.

Zdun, M.C. 1979: Continuous and differentiable iteration semigroups
Prace Naukowe Uniwersitetu Śląskiego W Katowicach Nr.308.

Zimmermann, G. 1978: Über die Existenz iterativer Wurzeln von Abbildungen.
Doctoral Dissertation, Marburg 1978.

Jürgen Weitkämper
Fachbereich Mathematik
Universität Marburg
Lahnberge
D-355 Marburg

ON EMBEDDING OF HOMEOMORPHISMS OF THE CIRCLE IN A CONTINUOUS FLOW

Marek Cezary Zdun

Let X be a subset of $\mathbb{C}$ or $\mathbb{R}$ and T be a function mapping X into X. A family of continuous functions $T^t:X \to X$, $t \in \mathbb{R}$ is said to be a flow or an iteration group (of T) if $T^t \circ T^s = T^{t+s}$ for $t,s \in \mathbb{R}$ (and $T^1 = T$). A flow $\{T^t\}$ is said to be continuous flow (abbreviated to c.f.) or continuous iteration group if for every $x \in X$ the mapping $t \to T^t(x)$ is continous (see[3,5]). A function T is said to be embeddable in a c.f. or has a c.f. if there exists a c.f. $\{T^t\}$ such that $T^1 = T$. A c.f. $\{T^t\}$ is said to be periodic with period d if $T^{t+d} = T^t$ for $t \in \mathbb{R}$ and $d > 0$ is the smallest number with this property.

Remark. If $\{T^t\}$ is a flow of a function T mapping X onto X, then $T^0(x) = x$ and each T^t is a homeomorphism from X onto X.

Proof. Suppose $x \in X$. Choose $y \in X$ such that $x = T(y)$. Then $T^0(x) = T^0(T(y)) = T(y) = x$. Moreover, we have $X = T^0[X] = T^t[T^{-t}[X]] \subset T^t[X] \subset X$ and $T^t \circ T^{-t}(x) = T^0(x) = x$ for $t \in \mathbb{R}$. Hence each T^t is a homeomorphism, since $(T^t)^{-1} = T^{-t}$ and all T^t are continuous.

Let $S^1 = \{z \in \mathbb{C}: |z| = 1\}$ be a circle with positive orientation. A subset $L \subset S^1$ is said to be an arc if it is connected, different from S^1 and card $L > 1$, i.e. $L = \{e^{it}: t \in \{a,b\}\}$, where $0 < |b-a| < 2\pi$.

Let $\{T^t\}$ be a c.f. Denote by $C(x)$ the orbit of x i.e.

$$C(x):= \{T^t(x): t \in \mathbb{R}\} \quad .$$

Moreover, put

$$A_T:= \{x \in S^1: T(x) = x\}$$

and

$$h_x(t):= T^t(x) \ , \quad t \in \mathbb{R} \ .$$

In the book 1 the following proposition is proved (Th.1.13, p.9).

Proposition 1. Let $\{T^t\}$ be a c.f. in a Hausdorff space and suppose $x \in X$ and $p(x):= \inf\{t > 0: T^t(x) = x\}$. Then

(a) $p(x) = 0$ iff $T^t(x) = x$ for every $t \in \mathbb{R}$,

(b) $0 < p(x) < \infty$ iff $C(x)$ is a Jordan curve,

(c) $p(x) = \infty$ iff the function $t \to T^t(x)$ is one-to-one.

Lemma 1. If $\{T^t\}$ is a c.f. of a homeomorphism $T:S^1 \to S^1$ and $\emptyset \neq A_T \neq S^1$, then for each $x \in A_T$ and $t \in \mathbb{R}$, $T^t(x) = x$.

Proof. Suppose $x_0 \in A_T$. Then the function $t \to T^t(x_0)$ is not one-to-one, since $T^0(x_0) = T^1(x_0)$. Moreover $C(x_0) \neq S^1$, since $C(x_0) \subset A_T$. In fact, if $y \in C(x_0)$ then $y = T^u(x_0)$ for an $u \in R$ and $T(y) = T \circ T^u(x_0) = T^u \circ T(x_0) = y$. Further by Prop.1 it follows that $T^t(x_0) = x_0$ for every $t \in R$. Assume $\emptyset \neq A_T \neq S^1$. Since A_T is a closed set we have the following decomposition

$$S^1 \setminus A_T = \bigcup_{\alpha \in M} L_\alpha \tag{1}$$

where L_α, $\alpha \in M$ are open disjoint arcs. This decomposition is unique. In this case we have the following

Theorem 1. Let $\{T^t\}$ be a c.f. of homeomorphisms $T:S^1 \to S^1$ and assume $\emptyset \neq A_T \neq S^1$. Moreover, let L_α, $\alpha \in M$ be the family of disjoint open arcs such that (1) holds. Then there exists a family of homeomorphisms $h_\alpha:\mathbb{R} \to L_\alpha$, such that

$$T^t(x) = \begin{cases} h_\alpha(t+h_\alpha^{-1}(x)), & x \in L_\alpha, \quad t \in \mathbb{R} \\ X, & X \in A_T, \quad t \in \mathbb{R}. \end{cases}$$

Conversely, formula (2) defines a c.f. on S^1.

Proof. Suppose $\alpha \in M$ and $t \in \mathbb{R}$. First we need to show that $T^t[L_\alpha] = L_\alpha$. Suppose $x \in L_\alpha$. Then clearly $T^t[C(x)] = C(x)$ and $C(x) \cap A_T = \emptyset$. The orbit $C(x)$ is an arc in S^1 as a continuous image of $\mathbb{R}$. Hence $C(x) \subset L_\alpha$, since $x \in C(x)$. Moreover $C(x) = T^t[C(x)] \subset T^t[L_\alpha]$, so $T^t[L_\alpha] \cap L_\alpha \neq \emptyset$. On the other hand it is easy to see that $T^t[L_\alpha]$ is an open arc disjoint with A_T. Therefore $T^t[L_\alpha] \subset L_\alpha$ and $L_\alpha = T^{-t}[T^t[L_\alpha]] \subset T^{-t}[L_\alpha]$. The last relations hold for every $t \in \mathbb{R}$ and $\alpha \in M$, so $T^t[L_\alpha] = L_\alpha$. Each arc L_α is homeomorphic to the interval (0,1). Let g_α, $\alpha \in M$ be homeomorphisms from L_α onto (0,1). Define $f_\alpha^t: = g_\alpha \circ T^t \circ g_\alpha^{-1}$. It is clear that $\{f^t\}$ is a c.f. on the interval (0,1). Then there exists a homeomorphism $a_\alpha:\mathbb{R} \xrightarrow[\text{onto}]{} (0,1)$, such that $f_\alpha^t(x) = a_\alpha(t+a_\alpha^{-1}(x))$ for $t \in \mathbb{R}$ (see [4,6] p.21 or [5] p.99). Put $h_\alpha: = g_\alpha^{-1} \circ a_\alpha$. Therefore $T^t(x) = h_\alpha(t+h_\alpha^{-1}(x))$ for $x \in L_\alpha$ and by L.1 $T^t(x) = x$ for $x \in A_T$ and $t \in \mathbb{R}$.

Conversely, it is easy to verify that (2) defines a c.f. on S^1. By Th.1 we get at once the following

Corollary 1. If $\{T^t\}$ is a c.f. of a homemorphism T and $\emptyset \neq A_T \neq S^1$, then for every $t \in \mathbb{R}$, $A_T t = A_T$.

Now consider the cases $A_T = \emptyset$ and $A_T = S^1$. In these cases for every fixed $t \in \mathbb{R}$ $A_T t = \emptyset$ for $A_T t = S^1$. In fact, suppose that $\emptyset \neq A_T t \neq S^1$. Define $G^s := T^{st}$ for $s \in \mathbb{R}$. $\{G^s\}$ is a c.f. of $G = T^t$ such that $G^{1/t} = T$. By Cor. 1 $A_T t = A_G 1 = A_G 1/t = A_T$, so $\emptyset \neq A_T \neq S^1$ but this contradicts our assumption.

After this remark define

$$d := \inf\{t > 0: \bigwedge_{z \in S^1} T^t(z) = z\} \qquad (\inf \emptyset := \infty) . \tag{3}$$

Lemma 2. Let $\{T^t\}$ be a c.f. of a homeomorphism T.

(a) If $A_T = \emptyset$, then $0 < d < \infty$.

(b) If $A_T = S^1$, then $d \leq 1$.

Proof. (a) Suppose $d = 0$. Then by Prop. 1(a) it follows that $T^t(z) = z$ for $t \in \mathbb{R}$ and $z \in S^1$, but this is a contradiction. Now suppose $d = \infty$. Then by Prop. 1(c) for each $z \in S^1$ the function $t \to T^t(z)$ is one-to-one. Therefore, for each $z \in S^1$ the orbit $C(z)$ is an open arc in S^1, since there does not exist any continuous one-to-one function from $\mathbb{R}$ onto S^1. It is clear that if $x \neq y$, then $C(x) \cap C(y) = \emptyset$ or $C(x) = C(y)$. Since S^1 is compact and $S^1 = \bigcup_{x \in S^1} C(x)$ there exists a finite covering of S^1 of orbits $C(x)$. On the other hand the finite sum of disjoint open arcs can not cover S^1. $d \neq \infty$.

(b) is trivial.

Lemma 3. Let $\{T^t\}$ be a c.f. of a homeomorphism T and $A_T = \emptyset$ or $A_T = S^1$.

(a) For every $z \in S^1$, we have $C(z) = S^1$ or for every $z \in S^1$, $C(z) = \{z\}$.

(b) If $d \neq 0$, then for every $z \in S^1$, $T^d(z) = z$ and the function $t \to T^t(z) =: h_z(t)$ is one-to-one in $\langle 0,d)$ and $S^1 = h_z\,[[0,d)]$.

Proof. (a) Suppose $T^t(z) \neq z$ for an $z \in S^1$ and a $t \in \mathbb{R}$. Then by L.2 $0 < d < \infty$. Furthermore by Prop. 1(b) $C(y) = S^1$ for every $y \in S^1$.

(b) Suppose that $z \in S^1$ and $T^u(z) = T^{u+s}(z)$ for some $u \geq 0$ and $s > 0$ such that $u+s < d$. Then $T^u(z) = T^u(T^s(z))$ and consequently $T^s(z) = z$. By the remark after Cor. 1 $T^s(z) \equiv z$ and this is impossible, since $s < d$. Thus h_z is one-to-one in $[0,d)$. By the continuity of h_z, $T^d(z) \equiv z$. Hence the function h_z is periodic with the period d. By the proven part (a), $S^1 = C(z) = h_z[\mathbb{R}] = h_z[[0,d)]$ for every $z \in S^1$. This completes the proof of the lemma.

One can easily show (c.f.[2] Chap.2, §3) that for every homeomorphism $T:S^1 \to S^1$ there exists a homeomorphism $f:\mathbb{R} \to \mathbb{R}$ such that

$$T(e^{2\pi ix}) = e^{2\pi if(x)}, \quad t \in \mathbb{R} \tag{3}$$

and $f(x+1) = f(x) + 1$, if f is increasing or $f(x+1) = f(x) - 1$, if f is decreasing.

We will say that the function f represents homeomorphism T. If f_1 and f_2 represent the same homeomorphism, then $f_1 = f_2 + k$ for a $k \in \mathbb{Z}$. Therefore there exists exactly one function f which represents T such that $0 \leq f(0) < 1$. If f is increasing, then we will say that homeomorphism T preserves orientation.

Remark. If T^t is a flow of homeomorphisms, then every T^t preserves orientation.

Proof. For every $t > 0$ $T^t = T^{t/2} \circ T^{t/2}$, thus T^t preserves orientation. Denote by

$$Q_a(z) := e^{2\pi i a} z \quad \text{for} \quad a \in R \quad \text{and} \quad z \in S^1 .$$

It is clear, that $\{Q_{at}\}$ is a periodic c.f. with period $d = 1/|a|$.

Theorem 2. If $\{T^t\}$ is a c.f. of a homeomorphism T and $A_T = \emptyset$ or $A_T = S^1$, then there exist an $a \in (-\infty,\infty)$ and an orientation preserving homeomorphisms $\Phi: S^1 \to S^1$ such that

$$T^t = \Phi^{-1} \circ Q_{at} \circ \Phi, \quad t \in \mathbb{R} . \tag{4}$$

Conversely, if $a \neq 0$, then (4) defines a continuous periodic flow with the period $d = 1/|a|$.

Proof. Suppose first that $d = 0$. Then by Prop. 1(a), $T^t(z) \equiv z$ for $t \in \mathbb{R}$. Therefore (4) holds for $a = 0$.

Now suppose that $d > 0$. Let $z_0 \in S^1$ and put $h := h_{z_0}|_{<0,d)}$. By L.3(b) the function h is one-to-one, continuous, $h(0) = h(d-)$ and $h[<0,d]] = S^1$. Denote by γ the function $t \to e^{2\pi it/d}$ restricted to $<0,d)$. It is clear that the function $\Phi := \gamma \circ h^{-1}$ is a homeomorphism from S^1 onto S^1, because S^1 is compact. Suppose $z \in S^1$. By L.3 there exists an $s \in <0,d)$ such that $z = h(s) = T^s(z_0)$. Hence

$$T^t(z) = T^t(T^s(z_0)) = T^{t+s}(z_0) = h_{z_0}(t+s) = h[(t+s)(\text{mod } d)] =$$

$$h[\gamma^{-1}(\gamma[(t+s)(\text{mod } d)])] = \Phi^{-1}(\gamma[(t+s)(\text{mod } d)]) = \Phi^{-1}(e^{2\pi i(t+s)/d}) =$$

$$\Phi^{-1}(e^{2\pi it/d}e^{2\pi is/d}) = \Phi^{-1}(Q_{t/d}(\gamma(s))) = \Phi^{-1}(Q_{t/d}(\gamma(h^{-1}(z)))) = \Phi^{-1}(Q_{t/d}(\Phi(z))) .$$

Thus we get (4) with $a = 1/d$.

Suppose that Φ does not preserve orientation. Put $\alpha(z) := \bar{z}$ for $z \in S^1$ and $\Psi := \alpha \circ \Phi$. It is clear that Ψ is a orientation preserving homeomorphism and $\alpha^{-1} = \alpha$ as well as $\alpha \circ Q_{-at} = Q_{at} \circ \alpha$. Hence

$$T^{-t} = \Phi^{-1} \circ Q_{-at} \circ \Phi = \Phi^{-1} \circ \alpha^{-1} \circ \alpha \circ Q_{-at} \circ \Phi = \Phi^{-1} \circ \alpha^{-1} \circ Q_{at} \circ \alpha \circ \Phi = \Psi^{-1} \circ Q_{at} \circ \Psi.$$

Thus $T^t(z) = \Psi^{-1}(Q_{-at}(\Psi(z)))$, where Ψ is a preserving-orientation homeomorphism of S^1.

Finally, it is easy to verify that (4) defines a c.f. with period $d = 1/|a|$.

Theorem 3. A homeomorphism $T: S^1 \to S^1$ is embeddable in a c.f. iff T preserves orientation and one of the following cases occurs:

1^0 $\emptyset \neq A_T \neq S^1$.

2^0 There exists a positive integer m such that $T^m(z) \equiv z$ in S^1.

3^0 For every $z \in S^1$, $\overline{\{T^n(z),\ n \in \mathbb{N}\}} = S^1$ i.e. T is a minimal homeomorphism.

Proof. For the time being we shall show only the necessity. First observe that in view of R.2 $T := T^1$ preserves orientation. Let $\{T^t\}$ be a c.f. of T and $A_T = \emptyset$ or $A_T = S^1$. In view of Th.2 there exists a homeomorphism $\Phi: S^1 \to S^1$ and an $a \in \mathbb{R}$ such that $T^t = \Phi^{-1} \circ Q_{at} \circ \Phi$. Hence $T = \Phi^{-1} \circ Q_a \circ \Phi$. If $a \in Q$, then there exists an $m \in \mathbb{N}$ such that $am \in Z$. Therefore $Q_{am}(z) \equiv z$ and consequently $T^m(z) \equiv z$. If $a \notin Q$, then for every $z \in S^1$ the set $\overline{\{Q_{am}(z),\ n \in \mathbb{N}\}} = \overline{\{e^{2\pi ian}z,\ n \in \mathbb{N}\}}$ is dense in S^1. Since Φ is a homeomorphism for each $z \in S^1$ $\overline{\{T^n(z),\ n \in \mathbb{N}\}} = \overline{\{\Phi^{-1}[Q_{an}(\Phi(z))],n \in \mathbb{N}\}} = \Phi^{-1}[\overline{\{Q_{na}(\Phi(z)),\ n \in N\}}] = \Phi^{-1}[S^1] = S^1$.

To prove the sufficient conditions we shall show some lemmas and theorems which give the general construction of all c.f.'s of T in each case separately.

First consider the case 1^0.

Theorem 4. If a homeomorphism $T: S^1 \to S^1$ preserves orientation and $\emptyset \neq A_T \neq S_1$, then T is embeddable in infinitely many c.f.'s depending on an arbitrary function.

Proof. To prove the theorem we shall show something more, namely we shall give the general construction of all c.f.'s of T.

Let $L_\alpha = (a_\alpha, b_\alpha)$, $\alpha \in M$ be a family of disjoint open arcs in S^1 such that $S^1 \setminus A_T = {}_{\alpha \in M} L$. Note that $a_\alpha, b_\alpha \in A_T$ for $\alpha \in M$.

Suppose $\alpha \in M$. It is clear that $T[L_\alpha]$ is an open arc and $T[L_\alpha] \cap A_T = \emptyset$. Hence it follows that there exist $\alpha, \beta \in M$ such that $T[L_\alpha] \subset L_\beta$. Since T is a homeomorphism and $a_\alpha, b_\alpha \in A_T$ it is clear that $a_\alpha, b_\alpha \in T[\overline{L_\alpha}] = \overline{T[L_\alpha]} \subset L_\beta$. On the other hand $L_\beta \cap A_T = \emptyset$. Hence it follows that either $L_\alpha = (a_\alpha, b_\alpha)$ or $L_\beta = (a_\alpha, b_\alpha)$. It is easy to see that the assumption that T preserves orientation rules out the case $L_\beta = (a_\alpha, b_\alpha)$. Therefore $T[L_\alpha] = L_\alpha$.

Let f_α be a homeomorphism from the arc L_α onto $(0,1)$. Define $g_\alpha := f_\alpha \circ T \circ f_\alpha^{-1}$. It is obvious that g_α is strictly increasing, $g(x) \neq x$ for $x \in (0,1)$, $g_\alpha(0+) = 0$ and $g_\alpha(1-) = 1$.

It is well-known (see [3], p.46) that the Abel equation

$$\gamma(g_\alpha(x)) = \gamma(x) + 1, \quad x \in (0,1) \tag{5}$$

has a continous and strictly monotonic solution $\gamma:(0,1) \xrightarrow[\text{onto}]{} \mathbb{R}$ depending on an arbitrary function. Moreover, the general form of all c.f.'s of g_α is the following

$$g_\alpha^t(x) := \gamma^{-1}(\gamma(x) + t), \qquad t \in R, \; x \in (0,1),$$

where γ is an arbitrary homeomorphism satisfying equation (5) (see[5], p.99). Hence and by Th.1 it follows that the formula

$$T^t(z) = \begin{cases} f_\alpha^{-1} \circ \gamma^{-1}(\gamma_\alpha \circ f_\alpha(z) + t), & z \in L_\alpha, \quad \alpha \in M, \quad t \in \mathbb{R} \\ z, & z \in A_T, \quad t \in \mathbb{R}, \end{cases}$$

where γ_α is an arbitrary homeomorphism from $(0,1)$ onto $\mathbb{R}$ satisfying equation (5) defines the general form of all c.f.'s of T. This completes the proof of the theorem.

By Th.2 we get easily the following

Theorem 5. The general form of all c.f.'s of the identity $T(z) \equiv z$ is the following

$$T^t = \Phi^{-1} \circ Q_{kt} \circ \Phi, \quad t \in R,$$

where $\Phi: S^1 \to S^1$ is an arbitrary homeomorphism and $k \quad \mathbb{Z}$.

Remark 3. If $A_T = \emptyset$, then the homeomorphism T preserves orientation.

Proof. Suppose that a homeormorphism T does not preserve orientation and f represents T. We may assume that $0 \leq f(0) < 1$. In this case f is decreasing and $f(x+1) = f(x)-1$, so $0 \leq f(1) < 1$. Then the continuity of f implies the existence of $t_1, t_2 \in [0,1)$ such that $f(t_1) = t_1 + 1$ and $f(t_2) = t_2$. Hence $T(x_k) = x_k$ for $x_k = e^{2\pi i t_k}$, so $A_T \neq \emptyset$, but this contradicts our assumption.

Lemma 4. If f represents a homeomorphism $F:S^1 \to S^1$ and g represents a homeomorphism $G:S^1 \to S^1$, then $f \circ g$ represents $F \circ G$.

Proof. Clear.

Lemma 5. If $\{T^t\}$ is a periodic c.f. with period $d > 0$, then there exists exactly one c.f. $\{f^t\}$ of real functions such that

$$f^t(x+1) = f^t(x) + 1 \qquad x,t \in \mathbb{R}, \tag{6}$$

$$f^{t+d}(x) = f^t(x) + 1 \qquad x,t \in \mathbb{R}, \tag{7}$$

$$T^t(e^{2\pi ix}) = e^{2\pi if^t(x)} \qquad x,t \in \mathbb{R}. \tag{8}$$

Conversely, if a c.f. $\{f^t\}$ satisfies (6) and (7), then (8) defines a periodic c.f. of homeomorphisms with period d.

Proof. Suppose $\{T^t\}$ is a c.f. of homeomorphisms with a period $d > 0$. By Th.1 we deduce that $A_T = \emptyset$ or $A_T = S^1$. However by Th.2 it follows that there exists a homeomorphism $\Phi:S^1 \to S^1$ such that

$$T^t(z) = \Phi^{-1}(e^{2\pi it/d}\Phi(z)) , \qquad t \in R, \quad z \in S^1. \tag{9}$$

Let a function $\phi:\mathbb{R} \to \mathbb{R}$ represent a homeomorphism Φ, i.e. ϕ is a homeomorphism such that $\phi(x+1) = \phi(x)+1$ or $\phi(x+1) = \phi(x) - 1$ and $\Phi(e^{2\pi ix}) = e^{2\pi i\phi(x)}$ for $x \in \mathbb{R}$. By L.4 it follows that $\Phi^{-1}(e^{2\pi ix}) = e^{2\pi i\phi^{-1}(x)}$. Hence by (9) we have

$$T^t(e^{2\pi ix})=\Phi^{-1}(e^{2\pi it/d}\Phi(e^{2\pi ix}))=\Phi^{-1}(e^{2\pi i(t/d+\phi(x))}) = e^{2\pi i\phi^{-1}(td+\phi(x))}.$$

Put $f^t(x): = \phi^{-1}(t/d + \phi(x))$, $t,x \in \mathbb{R}$.

It is clear that $\{f^t\}$ is a c.f. satisfying (6), (7) and (8). Note that $\{f^t\}$ is determined uniquely. In fact, let $\{g^t\}$ be a c.f. satisfying (6), (7) and (8). Then $e^{2\pi if^t(x)} = e^{2\pi ig^t(x)}$ for $t,x \in \mathbb{R}$. Therefore $f^t(x) = g^t(x) + k(t,x)$, where $k(t,x) \in \mathbb{Z}$.

The function $k:\mathbb{R}^2 \to \mathbb{Z}$ is constant as a difference of the continuous functions $(t,x) \to f^t(x)$ and $(t,x) \to g^t(x)$. On the other hand $f^0(x) = x = g^0(x)$, so $k \equiv 0$. Thus $f^t = g^t$ for $t \in \mathbb{R}$.

Conversely, if $\{f^t\}$ is a c.f. satisfying (6) and (7), then it is clear that formula (8) defines a c.f. with period d. Now we shall consider the case 2^0 of Th.3.

Lemma 6. Let a function $f:\mathbb{R} \to \mathbb{R}$ represent a homeomorphism $T:S^1 \to S^1$ such that $A_T = \emptyset$, $T^m(z) = z$ for an $m \in \mathbb{N}$ and m be the minimal positive integer with this property. If $0 \leq f(0) < 1$, then

$$x < f(x) < x + 1, \qquad x \in \mathbb{R} \tag{10}$$

and there exists a positive integer $q < m$ such that $(m,q) = 1$ and

$$f^m(x) = x + q, \quad x \in \mathbb{R}. \tag{11}$$

Proof. Note that the assumption $A_T = \emptyset$ implies that $e^{2\pi i f(x)} \neq e^{2\pi i x}$ for $x \in \mathbb{R}$. Hence $f(x) - x \in \mathbb{Z}$ for $x \in \mathbb{R}$. Further by the continuity of f and the assumption that $0 \leq f(0) < 1$ it follows that $0 < f(x) - x < 1$ for $x \in \mathbb{R}$. Thus inequality (10) holds. By L.4 we have $T^m(e^{2\pi i x}) = e^{2\pi i f^m(x)}$. Hence $e^{2\pi i f^m(x)} = e^{2\pi i x}$ and consequently $f^m(x) - x = :k(x) \in \mathbb{Z}$. It is clear that $k(x) \equiv q$ for a $q \in \mathbb{Z}$, since k is a continuous function with values in $\mathbb{Z}$. Thus we get (11). In view of (10) we get by the induction that $x < f^m(x) < x + m$. Hence, by (11) we get $x < x + q < x + m$. Thus $1 \leq q \leq m - 1$.

Put $j: = (m,q)$. Then there exist $m',q' \in \mathbb{N}$ such that $m = m'j$ and $q = q'j$. Suppose that there exists an x_0 such that $f^{m'}(x_0) < x_0 + q'$. Note that $f^{m'}$ is strictly increasing and $f^{m'}(x + q') = f^{m'}(x) + q'$. Hence it is easy to show by the induction that $f^{km'}(x_0) < x_0 + kq'$ for $k \in \mathbb{N}$. For $k = j$ the last inequality contradicts (10). Thus $f^{m'}(x) \geq x + q'$. If we suppose that there exists an x_0 such that $f^{m'}(x_0) > x_0 + q'$, then using the same argument as above we get also the contradiction to (10). Thus $f^{m'}(x) = x + q'$ for $x \in \mathbb{R}$. Therefore $T^{m'}(z) = z$ for $z \in S^1$. Hence according to the minimal property of the integer m we have that $m = m'$ and consequently $j = 1$. Thus we have proved that $(m,q) = 1$.

From elementary member theory we have the following

Remark 4. If $(q,m) = 1$ and $1 \leq q < m$, then there exists exactly one positive integer $p < m$ such that $pq = 1 (\mathrm{mod}\ m)$.

Lemma 7. Suppose p,q,m,n are integers such that $1 \leq q < m$, $(q,m) = 1$ and $pq = 1 + nm$. If $f(x+1) = f(x)+1$ and $f^m(x) = x+q$, for $x \in \mathbb{R}$, then the system of functional equations

$$\left.\begin{aligned} &\beta(f(x)) = \beta(x) + 1 \\ &\beta(x+1) = \beta(x) + m/q \end{aligned}\right\}, \quad x \in \mathbb{R} \tag{12}$$

is equivalent to the equation

$$\beta(f^p(x) - n) = \beta(x) + 1/q, \quad x \in \mathbb{R}. \tag{13}$$

Proof. If β satisfies (12), then $\beta(f^p(x) - k) = \beta(f^p(x)) - nm/q = \beta(x) + p - nm/q = \beta(x) + 1/q$ for $x \in \mathbb{R}$.

Put $g(x): = f^p(x) - n$. Note that $g^m(x) = f^{mp}(x) - mn = x + pq - mn = x+1$ for $x \in \mathbb{R}$ and $g^q(x) = f^{qp}(x) - qn = f^{1+mn}(x) - qn = f^{pm}(f(x)) - qn = f(x) + nq - qn = f(x)$.

Now, let β satisfy (13). Then we have

$$\beta(x+1) = \beta(g^m(x)) = \beta(x) + m/q, \quad \text{for} \quad x \in \mathbb{R} \text{ and}$$

$$\beta(f(x)) = \beta(g^0(x)) = \beta(x) + q/q = \beta(x) + 1 \quad \text{for} \quad x \in \mathbb{R}.$$

Let $T:S^1 \to S^1$ be orientation-preserving homeomorphism represented by a function f. The number $\alpha(T) \in [0,1)$ such that

$$\alpha(T) = \lim_{n\to\infty} \frac{f^n(x)}{n} \pmod 1, \quad x \in \mathbb{R}$$

is said to be the rotation number of T. This limit always exists and does not depend on x and f (see [2]). It is easy to show the following

Remark 5. (See [2], ch.2, §3). If T,U and Φ are preserving-orientation homeomorphisms and $T = \Phi^{-1} \circ U \circ \Phi$, then $\alpha(T) = \alpha(U)$.

Theorem 6. Suppose $T:S^1 \to S^1$ is orientation-preserving homeomorphism and there exists an integer $m > 1$ such that $T^m(z) \equiv z$ and $A_{T^k} \neq S^1$ for $1 \leq k < m$. Then T is embeddable in infinitely many c.f.'s depending on an arbitrary function. More precisely, if f represents T, $0 \leq f(0) < 1$, $q: = m\alpha(T)$ and p is an integer determined by Remark 4, then the general form of all c.f.'s of T is as follows:

$$T^t(e^{2\pi ix}) = e^{2\pi i\, \gamma^{-1}[t(\alpha(T)+k)m + \gamma(x)]}, \quad t,x \in \mathbb{R}, \tag{14}$$

where $k \in \mathbb{Z}$ and γ is an arbitrary continuous and strictly increasing solution of the Abel equation

$$\gamma(f^p(x) - \frac{pq-1}{m}) = \gamma(x) + 1, \quad x \in \mathbb{R} . \tag{15}$$

Proof. Let $\{T^t\}$ be a c.f. of T. It is clear that $A_T = \emptyset$. By Th.2 there exist a orientation-preserving homeomoprhism $\Phi:S^1 \to S^1$ and an $a \neq 0$ such that $T^t = \Phi^{-1} \circ$ $\circ\, Q_{ta} \circ \Phi$ for $t \in \mathbb{R}$. Therefore $T = \Phi^{-1} \circ Q_a \circ \Phi$ and moreover by R.5 $\alpha(T) = \alpha(Q_a) =$ $= a \pmod 1$, so $a = \alpha(T) + k$ for a $k \in \mathbb{Z}$. Put

$$U^t: = T^{\alpha(T)t/(\alpha(T)+k)}, \quad t \in \mathbb{R}.$$

Note that $U^1 = T^{\alpha(T)/(\alpha(T)+k)} = T^{1-k/(\alpha(T)+k)} = T^{1-k/a} = T^1 = T$, since $1/|a|$ is a period of $\{T^t\}$. It is also clear that $\{U^t\}$is a continuous periodic flow with period

$d = 1/\alpha(T) > 1$. Further by L.5 there exists a c.f. $\{f^t\}$ of real functions such that (6) and (7) hold and

$$U^t(e^{2\pi ix}) = e^{2\pi if^t(x)}, \quad x,t \in \mathbb{R}.$$

Hence it follows that $f: = f^1$ represents the homeomorphism T. Moreover in view of (6) $0 = f^0(0) < f^1(0) < f^d(0) = 1$, since the function $t \to f^t(0)$ is strictly increasing and $d > 1$. Thus $0 < f(0) < 1$. Further on account of L.6 there exists a positive integer $q < m$ such that $(m,q) = 1$ and (11) holds.

It is clear that there exists a homeomorphis $\beta:\mathbb{R} \xrightarrow[\text{onto}]{} \mathbb{R}$ such that

$$f^t(x) = \beta^{-1}(t + \beta(x)), \quad x,t \in \mathbb{R} \tag{18}$$

and $\beta(0) = 0$ (see [5], p.99). Hence

$$\beta(f(x)) = \beta(x) + 1, \quad x \in \mathbb{R}$$

since $f^1 = f$. Therefore

$$\beta(f^m(x)) = \beta(x) + m, \quad x \in \mathbb{R}$$

and by (11) we get

$$\beta(x + q) = \beta(x) + m , \quad x \in \mathbb{R} . \tag{19}$$

On the other hand by (6) we have

$$\beta^{-1}(t+\beta(x+1)) = \beta^{-1}(t+\beta(x)) + 1 , \quad x,t \in \mathbb{R}.$$

Inserting $x = 0$ in the last equality we get

$$\beta^{-1}(t + \beta(1)) = \beta^{-1}(t) + 1 , \quad t \in \mathbb{R}.$$

Therefore

$$\beta(x+1) = \beta(x) + \beta(1), \quad x \in \mathbb{R}$$

Hence

$$\beta(x+q) = \beta(x) + q\beta(1), \quad x \in \mathbb{R} .$$

Comparing the last equation with (19) we get that $\beta(1) = m/q$. Thus β satisfies the system of equations (12). By R.4 there exist positive integers $p < m$ and $n \in \mathbb{Z}$ such that $pq = 1 + nm$. Next in view of L.7, β satisfies equation (13). Hence the function $\gamma: = q\beta$ satisfies (15), since $n = \frac{pq-1}{m}$.

Directly by the definition of $\alpha(T)$ and (11) it follows that $\alpha(T) = q/m$. Hence according to (16), (17) and (18) we have

$$T^t(e^{2\pi ix}) = U^{(\alpha(T)+k)t/\alpha(T)}(e^{2\pi ix}) = e^{2\pi i\beta^{-1}(\frac{(\alpha(T)+k)t}{\alpha(T)} + \beta(x))} =$$

$$e^{2\pi i\,\gamma^{-1}(tq\frac{\alpha(T)+k}{\alpha(T)} + \gamma(x))} = e^{2\pi i\,\gamma^{-1}(t(\alpha(T)+k)m+\gamma(x))}, \qquad x,t \in \mathbb{R}.$$

Thus every c.f. of T is of the form (14), where γ is a homeomorphism satisfying (15).

Now we shall show that for every continuous and strictly increasing solution of (15) formula (14) defines a c.f. of T. First note that equation (15) has a continuous and strictly increasing solution depending on an arbitrary function (see [3], p.46). Let γ be an arbitrary strictly increasing and continuous solution of (15). It is clear that (14) defines a c.f. $\{T^t\}$. Note that the function $\beta = \gamma/q$ satisfies equation (13), where $n = \frac{pq-1}{m}$ and p,q are determined by L.6 and R.4. Of course $q = m\alpha(T)$. Hence it follows in view of L.7 that β satisfies the system of equations (12). Therefore

$$f(x) = \beta^{-1}(1 + \beta(x)), \qquad x \in \mathbb{R}$$

and

$$\beta^{-1}(x + 1/\alpha(T)) = \beta^{-1}(x) + 1, \qquad x \in \mathbb{R}.$$

Hence according to (14) we have

$$T^1(e^{2\pi ix}) = e^{2\pi i\,\beta^{-1}(\frac{\alpha(T)+k}{\alpha(T)} + \beta(x))} = e^{2\pi i\beta^{-1}(1+\beta(x) + k/\alpha(T))} =$$

$$= e^{2\pi i[\beta^{-1}(1+\beta(x))+k]} = e^{2\pi i\,\beta^{-1}(1+\beta(x))} = e^{2\pi if(x)} = T(e^{2\pi ix}).$$

Thus $T^1 = T$. This completes the proof of the theorem.

Finally we shall consider the case 3^0 of Th.3.

Theorem 7. Every minimal homeomorphism $T:S^1 \to S^1$ has exactly one c.f. with period $1/\alpha(T)$.

Proof. First observe that by R.3 it follows that T preserves orientation. Note also that $\alpha(T) \notin Q$. In fact, suppose $\alpha(T) \in Q$, then there exist an $x_0 \in S^1$ and an $m \in \mathbb{N}$ such that $T^m(x_0) = x_0$ (see [2], Ch.3, §3). Hence the homeomorphism T is not minimal, since $\{T^n(x_0), n \in \mathbb{N}\} = \{x_0, T(x_0), \ldots, T^{m-1}(x_0)\}$.

It is proved in the book [2]/Ch.3, §3, Th.3.1) that under our assumptions there exists an orientation-preserving homeomorphism $:S^1 \to S^1$ such that

$$\Phi \circ T = Q_{\alpha(T)} \circ \Phi \,.$$

It is clear that $T^t := \Phi^{-1} \circ Q_{t\alpha(T)} \circ \Phi$ is a c.f. of T with period $d = 1/\alpha(T)$. Now we shall show the uniqueness.

Let $\{G^t\}$ be a c.f. of T with the period $d = 1/\alpha(T) > 1$. On account of L.5 there exist c.f.'s $\{f^t\}$ and $\{g^t\}$ such that

$$T^t(e^{2\pi ix}) = e^{2\pi if^t(x)}, \quad G^t(e^{2\pi ix}) = e^{2\pi ig^t(x)}, \quad x,t \in \mathbb{R} \tag{20}$$

and

$$f^{t+d}(x) = f^t(x)+1, \quad g^{t+d}(x) = g^t(x)+1, \quad x,t \in \mathbb{R}. \tag{21}$$

Note that $0 = f^0(0) < f^1(0) < f^d(0) = 1$ and $0 = g^0(0) < g^1(0) < g^d(0) = 1$ since the functions $t \to f^t(0)$ and $t \to g^t(0)$ are strictly increasing. Hence it follows that $f^1 = g^1$, since f^1 and g^1 represent the same homeomorphism T. For c.f.'s $\{f^t\}$ and $\{g^t\}$ there exist homeomorphisms α and β from $\mathbb{R}$ onto $\mathbb{R}$ such that

$$f^t(x) = \alpha^{-1}(t+ (x)), \quad g^t(x) = \beta^{-1}(t+\beta(x)), \quad t,x \in \mathbb{R}, \tag{22}$$

$\alpha(0) = 0$ and $\beta(0) = 0$ (see [5]).
Put $\gamma := \alpha \circ \beta^{-1}$. From the equality $f^1 = g^1$ we have

$$\beta^{-1}(1 + \beta(x)) = \alpha^{-1}(1 + \alpha(x)), \quad x \in \mathbb{R}.$$

Then

$$\gamma(x+1) = \gamma(x) + 1, \quad x \in \mathbb{R}. \tag{23}$$

However by (21) and (22)

$$\alpha^{-1}(t+d+\alpha(x)) = \alpha^{-1}(t+\alpha(x)) + 1, \quad x,t \in \mathbb{R}$$

and

$$\beta^{-1}(t+d+\beta(x)) = \beta^{-1}(t+\beta(x)) + 1, \quad x,t \in R\,.$$

Inserting x = 0 in the last equalities we get $\beta^{-1}(t+d) = \beta^{-1}(t) + 1$ and $\alpha^{-1}(t+d) = \alpha^{-1}(t) + 1$, whence $\alpha(t+1) = \alpha(t) + d$ and consequently

$$\gamma(x+d) = \alpha(\beta^{-1}(x+d)) = \alpha(\beta^{-1}(x)+1) = \alpha(\beta^{-1}(x)) + d = \gamma(x) + d, \quad x \in \mathbb{R}.$$

Hence according to (23) we get

$$\gamma(x+n+kd) = \gamma(x+n) + kd = \gamma(x) + n + kd, \quad \text{for} \quad k,n \in \mathbb{Z}, \quad x \in \mathbb{R}\ .$$

Thus

$$\gamma(n + kd) = n + kd, \quad \text{for} \quad k,n \in \mathbb{Z}\ ,$$

since $\gamma(0) = 0$. Hence we deduce that $\gamma(x) = x$ for $x \in \mathbb{R}$, since γ is continuous and the set $\{n+kd; n,k \in \mathbb{Z}\}$ is dense in $\mathbb{R}$. Thus $\alpha = \beta$ and consequently by (22) and (20) $T^t = G^t$ for $t \in \mathbb{R}$. This complets the proof of the theorem.

Theorem 8. If $T:S^1 \to S^1$ is a minimal homeomorphism and $\{T^t\}$ is a c.f. of T with the period $1/\alpha(T)$ (which is uniquely determined by Th.7), then the general form of all c.f.'s of T is the following:

$$G^t = T^{t(\alpha(T)+k)/\alpha(T)} \quad , \quad t \in \mathbb{R}, \tag{24}$$

where $k \in \mathbb{Z}$, the period of $\{G^t\}$ is equal $1/(\alpha(T)+k)$.

Proof. Suppose $\{G^t\}$ is a c.f. of T. By Th.2 it follows that there exist an orientation-preserving homeomorphism $\Phi:S^1 \to S^1$ and an $a \neq 0$ such that $G^t = \Phi^{-1} \circ Q_{at} \circ Q$ for $t \in \mathbb{R}$. Hence $T = \Phi^{-1} \circ Q_a \circ \Phi$ and further we infer that $\alpha(T) = \alpha(Q_a) = a (\mathrm{mod}\ 1)$. Put $k: = a-\alpha(T)$ and $U^t: = G^{t\alpha(T)/(\alpha(T)+k)}$ $t \in \mathbb{R}$. It is readily verified that $\{U^t\}$ is a c.f. of T with the period $1/\alpha(T)$. Hence according to Th.7 $U^t = T^t$ for $t \in \mathbb{R}$. Thus we get (24).

It is clear that $\{G^t\}$ defined by (24) is a c.f. with the period $1/(\alpha(T)+k)$.

REFERENCES

[1] Beck, A.: Continuous flows in the plane. Grundlehren 201, Springer Verlag, Berlin, Heidelberg, New York, 1974.

[2] Cornfeld, I.P., Fomin, S.V., Sinai, Y.G.: Ergodic theory, Grundlehren 245, Springer Verlag, Berlin, Heidelberg, New York, 1982.

[3] Kuczma, M.: Functional equations in a single variable. Monografie Mat. PWN Warszawa, 1968.

[4] Midura, S.: Sur les solutions de l'equation de translation, Aequationes Math. 1, 77-84 (1968).

[5] Targonski, Gy.: Topics in iteration theory. Vandenhoeck and Ruprecht in Göttingen und Zürich, 1981.

[6] Zdun, M.C.: Continuous and differentiable iteration semigroups. Prace Naukowe Uniwersytetu Śląskiego w Katowicach Nr. 308 (1979).

M.C. Zdun

Institut of Mathematics

Pedagogical University, Kraków

ul. Podchorażych 2

PL-30-084 Kraków, Poland

Vol. 1008: Algebraic Geometry. Proceedings, 1981. Edited by J. Dolgachev. V, 138 pages. 1983.

Vol. 1009: T. A. Chapman, Controlled Simple Homotopy Theory and Applications. III, 94 pages. 1983.

Vol. 1010: J.-E. Dies, Chaînes de Markov sur les permutations. IX, 226 pages. 1983.

Vol. 1011: J.M. Sigal. Scattering Theory for Many-Body Quantum Mechanical Systems. IV, 132 pages. 1983.

Vol. 1012: S. Kantorovitz, Spectral Theory of Banach Space Operators. V, 179 pages. 1983.

Vol. 1013: Complex Analysis - Fifth Romanian-Finnish Seminar. Part 1. Proceedings, 1981. Edited by C. Andreian Cazacu, N. Boboc, M. Jurchescu and I. Suciu. XX, 393 pages. 1983.

Vol. 1014: Complex Analysis - Fifth Romanian-Finnish Seminar. Part 2. Proceedings, 1981. Edited by C. Andreian Cazacu, N. Boboc, M. Jurchescu and I. Suciu. XX, 334 pages. 1983.

Vol. 1015: Equations différentielles et systèmes de Pfaff dans le champ complexe - II. Seminar. Edited by R. Gérard et J.P. Ramis. V, 411 pages. 1983.

Vol. 1016: Algebraic Geometry. Proceedings, 1982. Edited by M. Raynaud and T. Shioda. VIII, 528 pages. 1983.

Vol. 1017: Equadiff 82. Proceedings, 1982. Edited by H. W. Knobloch and K. Schmitt. XXIII, 666 pages. 1983.

Vol. 1018: Graph Theory, Łagów 1981. Proceedings, 1981. Edited by M. Borowiecki, J. W. Kennedy and M. M. Sysło. X, 289 pages. 1983.

Vol. 1019: Cabal Seminar 79-81. Proceedings, 1979-81. Edited by A. S. Kechris, D. A. Martin and Y. N. Moschovakis. V, 284 pages. 1983.

Vol. 1020: Non Commutative Harmonic Analysis and Lie Groups. Proceedings, 1982. Edited by J. Carmona and M. Vergne. V, 187 pages. 1983.

Vol. 1021: Probability Theory and Mathematical Statistics. Proceedings, 1982. Edited by K. Itô and J.V. Prokhorov. VIII, 747 pages. 1983.

Vol. 1022: G. Gentili, S. Salamon and J.-P. Vigué. Geometry Seminar "Luigi Bianchi", 1982. Edited by E. Vesentini. VI, 177 pages. 1983.

Vol. 1023: S. McAdam, Asymptotic Prime Divisors. IX, 118 pages. 1983.

Vol. 1024: Lie Group Representations I. Proceedings, 1982-1983. Edited by R. Herb, R. Lipsman and J. Rosenberg. IX, 369 pages. 1983.

Vol. 1025: D. Tanré, Homotopie Rationnelle: Modèles de Chen, Quillen, Sullivan. X, 211 pages. 1983.

Vol. 1026: W. Plesken, Group Rings of Finite Groups Over p-adic Integers. V, 151 pages. 1983.

Vol. 1027: M. Hasumi, Hardy Classes on Infinitely Connected Riemann Surfaces. XII, 280 pages. 1983.

Vol. 1028: Séminaire d'Analyse P. Lelong - P. Dolbeault - H. Skoda. Années 1981/1983. Edité par P. Lelong, P. Dolbeault et H. Skoda. VIII, 328 pages. 1983.

Vol. 1029: Séminaire d'Algèbre Paul Dubreil et Marie-Paule Malliavin. Proceedings, 1982. Edité par M.-P. Malliavin. V, 339 pages. 1983.

Vol. 1030: U. Christian, Selberg's Zeta-, L-, and Eisensteinseries. XII, 196 pages. 1983.

Vol. 1031: Dynamics and Processes. Proceedings, 1981. Edited by Ph. Blanchard and L. Streit. IX, 213 pages. 1983.

Vol. 1032: Ordinary Differential Equations and Operators. Proceedings, 1982. Edited by W. N. Everitt and R. T. Lewis. XV, 521 pages. 1983.

Vol. 1033: Measure Theory and its Applications. Proceedings, 1982. Edited by J. M. Belley, J. Dubois and P. Morales. XV, 317 pages. 1983.

Vol. 1034: J. Musielak, Orlicz Spaces and Modular Spaces. V, 222 pages. 1983.

Vol. 1035: The Mathematics and Physics of Disordered Media. Proceedings, 1983. Edited by B. D. Hughes and B. W. Ninham. VII, 432 pages. 1983.

Vol. 1036: Combinatorial Mathematics X. Proceedings, 1982. Edited by L. R. A. Casse. XI, 419 pages. 1983.

Vol. 1037: Non-linear Partial Differential Operators and Quantization Procedures. Proceedings, 1981. Edited by S. I. Andersson and H.-D. Doebner. VII, 334 pages. 1983.

Vol. 1038: F. Borceux, G. Van den Bossche, Algebra in a Localic Topos with Applications to Ring Theory. IX, 240 pages. 1983.

Vol. 1039: Analytic Functions, Błażejewko 1982. Proceedings. Edited by J. Ławrynowicz. X, 494 pages. 1983

Vol. 1040: A. Good, Local Analysis of Selberg's Trace Formula. III, 128 pages. 1983.

Vol. 1041: Lie Group Representations II. Proceedings 1982-1983. Edited by R. Herb, S. Kudla, R. Lipsman and J. Rosenberg. IX, 340 pages. 1984.

Vol. 1042: A. Gut, K. D. Schmidt, Amarts and Set Function Processes. III, 258 pages. 1983.

Vol. 1043: Linear and Complex Analysis Problem Book. Edited by V. P. Havin, S. V. Hruščëv and N. K. Nikol'skii. XVIII, 721 pages. 1984.

Vol. 1044: E. Gekeler, Discretization Methods for Stable Initial Value Problems. VIII, 201 pages. 1984.

Vol. 1045: Differential Geometry. Proceedings, 1982. Edited by A. M. Naveira. VIII, 194 pages. 1984.

Vol. 1046: Algebraic K-Theory, Number Theory, Geometry and Analysis. Proceedings, 1982. Edited by A. Bak. IX, 464 pages. 1984.

Vol. 1047: Fluid Dynamics. Seminar, 1982. Edited by H. Beirão da Veiga. VII, 193 pages. 1984.

Vol. 1048: Kinetic Theories and the Boltzmann Equation. Seminar, 1981. Edited by C. Cercignani. VII, 248 pages. 1984.

Vol. 1049: B. Iochum, Cônes autopolaires et algèbres de Jordan. VI, 247 pages. 1984.

Vol. 1050: A. Prestel, P. Roquette, Formally p-adic Fields. V, 167 pages. 1984.

Vol. 1051: Algebraic Topology, Aarhus 1982. Proceedings. Edited by I. Madsen and B. Oliver. X, 665 pages. 1984.

Vol. 1052: Number Theory. Seminar, 1982. Edited by D.V. Chudnovsky, G.V. Chudnovsky, H. Cohn and M.B. Nathanson. V, 309 pages. 1984.

Vol. 1053: P. Hilton, Nilpotente Gruppen und nilpotente Räume. V, 221 pages. 1984.

Vol. 1054: V. Thomée, Galerkin Finite Element Methods for Parabolic Problems. VII, 237 pages. 1984.

Vol. 1055: Quantum Probability and Applications to the Quantum Theory of Irreversible Processes. Proceedings, 1982. Edited by L. Accardi, A. Frigerio and V. Gorini. VI, 411 pages. 1984.

Vol. 1056: Algebraic Geometry. Bucharest 1982. Proceedings, 1982. Edited by L. Bădescu and D. Popescu. VII, 380 pages. 1984.

Vol. 1057: Bifurcation Theory and Applications. Seminar, 1983. Edited by L. Salvadori. VII, 233 pages. 1984.

Vol. 1058: B. Aulbach, Continuous and Discrete Dynamics near Manifolds of Equilibria. IX, 142 pages. 1984.

Vol. 1059: Séminaire de Probabilités XVIII, 1982/83. Proceedings. Edité par J. Azéma et M. Yor. IV, 518 pages. 1984.

Vol. 1060: Topology. Proceedings, 1982. Edited by L. D. Faddeev and A.A. Mal'cev. VI, 389 pages. 1984.

Vol. 1061: Séminaire de Théorie du Potentiel. Paris, No. 7. Proceedings. Directeurs: M. Brelot, G. Choquet et J. Deny. Rédacteurs: F. Hirsch et G. Mokobodzki. IV, 281 pages. 1984.

Vol. 1062: J. Jost, Harmonic Maps Between Surfaces. X, 133 pages. 1984.

Vol. 1063: Orienting Polymers. Proceedings, 1983. Edited by J.L. Ericksen. VII, 166 pages. 1984.

Vol. 1064: Probability Measures on Groups VII. Proceedings, 1983. Edited by H. Heyer. X, 588 pages. 1984.

Vol. 1065: A. Cuyt, Padé Approximants for Operators: Theory and Applications. IX, 138 pages. 1984.

Vol. 1066: Numerical Analysis. Proceedings, 1983. Edited by D.F. Griffiths. XI, 275 pages. 1984.

Vol. 1067: Yasuo Okuyama, Absolute Summability of Fourier Series and Orthogonal Series. VI, 118 pages. 1984.

Vol. 1068: Number Theory, Noordwijkerhout 1983. Proceedings. Edited by H. Jager. V, 296 pages. 1984.

Vol. 1069: M. Kreck, Bordism of Diffeomorphisms and Related Topics. III, 144 pages. 1984.

Vol. 1070: Interpolation Spaces and Allied Topics in Analysis. Proceedings, 1983. Edited by M. Cwikel and J. Peetre. III, 239 pages. 1984.

Vol. 1071: Padé Approximation and its Applications, Bad Honnef 1983. Prodeedings. Edited by H. Werner and H.J. Bünger. VI, 264 pages. 1984.

Vol. 1072: F. Rothe, Global Solutions of Reaction-Diffusion Systems. V, 216 pages. 1984.

Vol. 1073: Graph Theory, Singapore 1983. Proceedings. Edited by K.M. Koh and H.P. Yap. XIII, 335 pages. 1984.

Vol. 1074: E.W. Stredulinsky, Weighted Inequalities and Degenerate Elliptic Partial Differential Equations. III, 143 pages. 1984.

Vol. 1075: H. Majima, Asymptotic Analysis for Integrable Connections with Irregular Singular Points. IX, 159 pages. 1984.

Vol. 1076: Infinite-Dimensional Systems. Proceedings, 1983. Edited by F. Kappel and W. Schappacher. VII, 278 pages. 1984.

Vol. 1077: Lie Group Representations III. Proceedings, 1982–1983. Edited by R. Herb, R. Johnson, R. Lipsman, J. Rosenberg. XI, 454 pages. 1984.

Vol. 1078: A.J.E.M. Janssen, P. van der Steen, Integration Theory. V, 224 pages. 1984.

Vol. 1079: W. Ruppert. Compact Semitopological Semigroups: An Intrinsic Theory. V, 260 pages. 1984

Vol. 1080: Probability Theory on Vector Spaces III. Proceedings, 1983. Edited by D. Szynal and A. Weron. V, 373 pages. 1984.

Vol. 1081: D. Benson, Modular Representation Theory: New Trends and Methods. XI, 231 pages. 1984.

Vol. 1082: C.-G. Schmidt, Arithmetik Abelscher Varietäten mit komplexer Multiplikation. X, 96 Seiten. 1984.

Vol. 1083: D. Bump, Automorphic Forms on GL (3,IR). XI, 184 pages. 1984.

Vol. 1084: D. Kletzing, Structure and Representations of Q-Groups. VI, 290 pages. 1984.

Vol. 1085: G.K. Immink, Asymptotics of Analytic Difference Equations. V, 134 pages. 1984.

Vol. 1086: Sensitivity of Functionals with Applications to Engineering Sciences. Proceedings, 1983. Edited by V. Komkov. V, 130 pages. 1984

Vol. 1087: W. Narkiewicz, Uniform Distribution of Sequences of Integers in Residue Classes. VIII, 125 pages. 1984.

Vol. 1088: A.V. Kakosyan, L.B. Klebanov, J.A. Melamed, Characterization of Distributions by the Method of Intensively Monotone Operators. X, 175 pages. 1984.

Vol. 1089: Measure Theory, Oberwolfach 1983. Proceedings. Edited by D. Kölzow and D. Maharam-Stone. XIII, 327 pages. 1984.

Vol. 1090: Differential Geometry of Submanifolds. Pro 1984. Edited by K. Kenmotsu. VI, 132 pages. 1984.

Vol. 1091: Multifunctions and Integrands. Proceedings, 19 by G. Salinetti. V, 234 pages. 1984.

Vol. 1092: Complete Intersections. Seminar, 1983. Edited by and R. Strano. VII, 299 pages. 1984.

Vol. 1093: A. Prestel, Lectures on Formally Real Fields. XI, 1 1984.

Vol. 1094: Analyse Complexe. Proceedings, 1983. Edité pa R. Gay et Nguyen Thanh Van. IX, 184 pages. 1984.

Vol. 1095: Stochastic Analysis and Applications. Proceedin Edited by A. Truman and D. Williams. V, 199 pages. 1984.

Vol. 1096: Théorie du Potentiel. Proceedings, 1983. Edit Mokobodzki et D. Pinchon. IX, 601 pages. 1984.

Vol. 1097: R.M. Dudley, H. Kunita, F. Ledrappier, École Probabilités de Saint-Flour XII – 1982. Edité par P.L. Henn 396 pages. 1984.

Vol. 1098: Groups – Korea 1983. Proceedings. Edited by and B.H. Neumann. VII, 183 pages. 1984.

Vol. 1099: C.M. Ringel, Tame Algebras and Integral Quadrati XIII, 376 pages. 1984.

Vol. 1100: V. Ivrii, Precise Spectral Asymptotics for Elliptic O Acting in Fiberings over Manifolds with Boundary. V, 23 1984.

Vol. 1101: V. Cossart, J. Giraud, U. Orbanz, Resolution of Singularities. VII, 132 pages. 1984.

Vol. 1102: A. Verona, Stratified Mappings – Structure and lability. IX, 160 pages. 1984.

Vol. 1103: Models and Sets. Proceedings, Logic Colloquiur Part I. Edited by G.H. Müller and M.M. Richter. VIII, 484 page

Vol. 1104: Computation and Proof Theory. Proceedings Colloquium, 1983, Part II. Edited by M.M. Richter, E. Börger, V schelp, B. Schinzel and W. Thomas. VIII, 475 pages. 1984.

Vol. 1105: Rational Approximation and Interpolation. Proce 1983. Edited by P.R. Graves-Morris, E.B. Saff and R.S. XII, 528 pages. 1984.

Vol. 1106: C.T. Chong, Techniques of Admissible Recursion IX, 214 pages. 1984.

Vol. 1107: Nonlinear Analysis and Optimization. Proceedings Edited by C. Vinti. V, 224 pages. 1984.

Vol. 1108: Global Analysis – Studies and Applications I. E Yu.G. Borisovich and Yu.E. Gliklikh. V, 301 pages. 1984.

Vol. 1109: Stochastic Aspects of Classical and Quantum S Proceedings, 1983. Edited by S. Albeverio, P. Combe and M. S Collin. IX, 227 pages. 1985.

Vol. 1110: R. Jajte, Strong Limit Theorems in Non-Comn Probability. VI, 152 pages. 1985.

Vol. 1111: Arbeitstagung Bonn 1984. Proceedings. Edited by bruch, J. Schwermer and S. Suter. V, 481 pages. 1985.

Vol. 1112: Products of Conjugacy Classes in Groups. Ed Z. Arad and M. Herzog. V, 244 pages. 1985.

Vol. 1113: P. Antosik, C. Swartz, Matrix Methods in Ana 114 pages. 1985.

Vol. 1114: Zahlentheoretische Analysis. Seminar. Herausg von E. Hlawka. V, 157 Seiten. 1985.

Vol. 1115: J. Moulin Ollagnier, Ergodic Theory and Statistical M ics. VI, 147 pages. 1985.

Vol. 1116: S. Stolz, Hochzusammenhängende Mannigfalti und ihre Ränder. XXIII, 134 Seiten. 1985.